Advances in Hemostasis and Thrombosis

Ettore Majorana International Science Series

Series Editor:
Antonino Zichichi
European Physical Society
Geneva, Switzerland

(LIFE SCIENCES)

A Continuation Order Plan is available for this series. A continuation order will bring delivery of each new volume immediately upon publication. Volumes are billed only upon actual shipment. For further information please contact the publisher.

Advances in Hemostasis and Thrombosis

Edited by

A. Cajozzo
R. Perricone
P. Di Marco
and

P. Palazzolo

University of Palermo
Palermo, Italy

Plenum Press • New York and London

Library of Congress Cataloging in Publication Data

International School of Medical Sciences (22nd: 1983: Ettore Majorana Center for
  Scientific Culture)
  Advances in hemostasis and thrombosis.

  (Ettore Majorana international science series. Life sciences; v. 20)
  "Proceedings of the International School of Medical Sciences 22nd course ... held
April 20–26, 1983, at the Ettore Majorana Center for Scientific Culture, Erice, Sicily,
Italy"—T.p. verso.
  Includes bibliographies and index.
  1. Blood—Coagulation, Disorders of—Congresses. 2. Blood—Coagulation, Disorders
of—Complications and sequelae—Congresses. 3. Thrombosis—Etiology—Congresses.
4. Hemostasis—Congresses. I. Cajozzo, A. II. Title. III. Series: Ettore Majorana in-
ternational science series. Life sciences; 20. [DNLM: 1. Blood Coagulation Disorders—
congresses. 2. Hemostasis—congresses. 3. Thrombosis—congresses.
WI ET721M v. 20 / WH 310 I597 1983a]
RC647.C55I58 1983   616.1'57   85-9264

ISBN 978-1-4615-9426-0   ISBN 978-1-4615-9424-6 (eBook)

DOI 10.1007/978-1-4615-9424-6

Proceedings of the International School of Medical Sciences
22nd course on Advances in Hemostasis and Thrombosis, held April 20–26, 1983,
at the Ettore Majorana Center for Scientific Culture, Erice, Sicily, Italy

©1985 Plenum Press, New York
Softcover reprint of the hardcover 1st edition 1985

A Division of Plenum Publishing Corporation
233 Spring Street, New York, N.Y. 10013

PREFACE

A greater knowledge of the basic molecular mechanisms which
regulate platelet function, coagulation, the fibrinolytic system
and their interaction with the vessel wall has allowed a better
understanding of the numerous aspects involved in the pathogenesis
of thrombosis and the regulation of hemostasis.

Thrombosis represents today one of the causes of death in
the Western hemisphere. Arterial thrombosis may lead to myocardial
infarction, cerebral and peripheral vascular diseases. Venous
thrombosis, on the other hand, is a complication of heart disease,
surgery, extensive trauma, several obstetric pathologies, and may,
at times, develop in apparently healthy subjects.

These complex problems are dealt with within the content of
"Advances in Hemostasis and Thrombosis." The present papers
contain up-to-date information on hemostasis, the role of
prostaglandins, fibrinolytic coagulation, rheology, and modern
therapeutic approaches along with their clinical application.

This book, therefore, offers the reader the opportunity to
profit from a rare encounter, an interdisciplinary cooperation,
with scientific contributions from numerous countries.

A. Cajozzo
R. Perricone
P. Di Marco
P. Palazzolo

CONTENTS

HEMOSTASIS: A REVIEW

Angelo Baserga

Università di Ferrara
Clinica Medica
Ferrara, Italy

INTRODUCTION

The aim of this lecture is to examine the development of our
knowledge of hemostasis. A chronological table of the main dates
of the history of coagulation and hemorrhagic diseases can be
found in my 1950 treatise "Le Malattie Emorragiche." A summary
of the history of coagulation can be found in various publications:
two chapters by Spaet and Ratnoff in Wintrobe's "Blood Pure and
Eloquent," and in a report presented by De Nicola in 1978 at the
International Congress of Bologna on thrombosis (regarding the
years 1935 to 1965). A summary of the history of platelets can
also be found in a 1981 publication presented at Bergamo in honor
of Bizzozzero's centenary. Therefore, a further analysis of the
development of such knowledge is unnecessary.

The various stages of our knowledge can be divided as follows:
1) forerunners - from Hewson (1771) to Alexander Schmidt (1876),
2) early scientific achievements - from Schmidt to Morawitz (1905),
3) standstill (also deviation and regression) stage - from 1905 to
1935, 4) renaissance stage (named by Quick) - from 1935 to 1960c.
(in 1954, Roman numerals were used for the plasma clotting factor
nomenclature), and lastly 5) the present stage of physiochemical
developments.

REGRESSION OF THE KNOWLEDGE OF HEMOSTASIS

In the present report, we shall discuss the events which led
to a slackening or to a regression of the knowledge of hemostasis.
At times, this was due to the experts' support of certain theories
(Bacon's so-called "idola theatri"). As a matter of fact, numerous
French students were misled by Hayem's statement that platelets
were precursors of red cells (Hématoblaste). In the above mentioned
book, Ratnoff analyzes the psychological and environmental conditions
which led to the acceptance of Hayem's erroneous opinions despite
contrasting findings. Of course, this was due to Hayem's authority
and influence.

Another example is Howell's theory (1911), which stated that
circulating blood did not clot because it contained inhibitors
that were normally joined to prothrombin. This theory overshadowed
Morawitz's previous correct concept. In this case, the discovery
of heparin by MacLeod, which took place five years later in Howell's
laboratory, seemed to give experimental evidence to Howell's hypothesis. Alternatively, the so-called "scientific fashions" (those
which Bacon called "idola fori") misled the students. Thus in the
20's, when enthusiasm for colloidal theories was greatest, they
wanted to refer the whole clotting process to a mere sol-gel
transformation.

UNUSED FINDINGS

We shall also discuss how adequately ascertained findings
unaccountably failed to be utilized. In fact, the discovery of
the anti-hemophilic globulin (Patek and Taylor, 1937) was queerly
not taken into account in the coagulation schemes for a whole
decade. When we review the old literature, it is interesting to
note that various forerunning ideas failed to be followed for
several decades. I shall site an instance which I think has not
been previously reported by any other hematology historian. The
French clinician Andral, author of the first book on pathologic
hematology (1843), was also the first to start a logical study of
hemorrhagic diseases, "Je crois pouvoir conclure qu'il y a ef-
fectivement un certain ombre d'hémorrhagies...qui sont le résultat
d'une modification dans la composition du sang; cette modification
consiste dans une diminution relative ou absolue de la fibrine"
(p. 108). Since then, the concept of hypo- and afibrinogenemic

syndromes has been established and has progressively evolved. On
the other hand, Andral went on stating, "au nombre des causes des
hémorrhagies, un état de dissolution du sang. Or qu'est-ce-que
cette dissolution, si ce n'est un état dans le quel le sang contient
moins de fibrine que de coutume?" At this point, the concept of
fibrinolytic conditions was established. However, Andral followed
only the former and not the latter aspect of his intuitions. Ninety
years went by before fibrinolytic syndromes were studied, using
Reimann's (1941) description of thrombolytic purpura. Allow me
to mention that the first treatise in which a specific chapter was
dedicated to the disease was written by me and De Nicola in 1950.

NON-MEDICAL CONTRIBUTIONS TO HEMOSTATIC KNOWLEDGE

The present report will also deal with the contributions that
fields alien to medical science made to hemostatic knowledge.
Dam's study of poultry led to the discovery of vitamin K. The
Dakota cattle breeders' observation of spontaneous bleeding in
cattle fed with sweet clover led to the identification of dicumarol.
The students of textile fibers interpreted the fibrin formation
as a polymerization process. Presently, the most advanced studies
of structural chemical analyses lead, for example, to the identification of the entire structure of prothrombin. However, one must
remember that some of the most important observations were made
by students who were working with poor means. Paul Owren identified
factor V and began research on the new plasma clotting factors in
a little Norwegian hospital that had been isolated during World
War II. Owren's very intelligent study was based on only one case.
In 1944, Pavlovski in Buenos Aires began the identification of the
two hemophilic patterns (A and B), using the mere mixture of blood
samples drawn from different patients.

HEMATOLOGICAL CONTRIBUTIONS TO OTHER FIELDS

Hemostatic studies have been extended to extra-hemorrhagic
fields. Quick's forerunning statement (1943), "from the point
of view of economy it seems unlikely that the body would utilize
as highly perfect mechanism as the coagulation of the blood solely
for the purpose of preventing hemorrhage" (p. 7 of his textbook).
This concept held true, for example, for therapeutic applications.
The vast field of anticoagulants stemmed from studies of heparin
and dicumarols. Numerous well known advantages derived from

postoperative prophylaxis: myocardial infarction management,
thrombophlebitic therapy, and so on. Conversely, iatrogenic
pathology also took place because of erroneous administration.
The dicumarol-induced hemorrhagic syndrome was cynically defined
as "a beast disease transplanted into men by beast-like doctors."
The study of platelet aggregation has also led to the wide field
of cardiovascular therapy, i.e., the platelet aggregation agents.
The knowledge of thromboxanes and various prostaglandins stemmed
from the study of platelets. Furthermore, several points of
fibrin pathology have shed some light on aspects of rheumatic,
collagen enoplastic affections, and so on.

ITALIAN SCIENTISTS' CONTRIBUTIONS TO HEMATOLOGY

I would also like to mention some of the contributions made
by Italian scientists to this field. In addition to Bizzozzero's
paramount research in 1882 on the platelets' role in blood clotting,
one must mention Sabbatani's work in the early 20th century on
the relationship between calcium ions and anticoagulants. One
must also mention Bizzozzero's discovery in 1868 of the "cellule
giganti con nucleo in gemmazione" (i.e., giant cells with budding
nucleus), which correspond to megakariocytes. Unfortunately,
such cells failed to be related to platelets for a long time.
Di Guglielmo's "Le Porpore Emorragiche" in 1926 and Samek's
treatise in 1930 are the first two books in world medical literature
specifically dedicated to hemorrhagic disease. Besides the research
carried out by our school, it is worthy to mention the important
role Italian students had in making evident the existence of
abnormal fibrinogens.

FUTURE OF HEMOSTATIC STUDIES

A review of the development of hemostatic values should not
only deal with the past, but also concern itself with the future.
It is likely that genetics will assume a greater importance in
the following years. Our school has constantly shown great interest
in cytogenetics. We shall, therefore, mostly concentrate on the
early achievements of chromosomal mapping, as well as on different
chromosomal patterns present in hemorrhagic diseases such as
hemophilia. By looking towards the future, we shall perhaps be
able to correct these problems through the use of genetic engineer-
ing techniques, or be able to produce plasma clotting factors

through synthesis, hybridoms, and gene transplants. However, it is difficult and dangerous to prophesize.

In conclusion, I would like to emphasize the importance that the early meeting of coagulationists had in accelerating the development of this field. I would like to mention the five conferences of the Macy Foundation of New York (1948-52). These abstracts show the vivacity and the occasional harshness of the opposing opinions (made famous by Quick's jeer, "some brains would be more useful as thromboplastin!"). The congress held at Montreux in 1949 by the European Society of Hematology, organized by Feissly, Fonion Fiehrer, Jorpes, Koller, Nolf, De Nicola, and myself, is worthy of mention. The former began the conferences on blood clotting while the latter led to the onset of specialized societies in hemostatic studies. Lastly, the famous Balducci Symposia presided by Roskam (Madrid, 1955), Seteasis (1957), Formentor (1960) renewed interest in hemostatic vascular factors and the relation-ship between endothelial wall and coagulation.

Let us hope that the present meeting will be equally creative and stimulating.

PLATELETS, ENDOTHELIUM AND

VESSEL INJURY

J.F. Mustard

McMaster University
Hamilton, Ontario
Canada - L8N 3Z5

INTRODUCTION

This subject includes platelet interaction with the normal
vessel wall, injured normal vessels, and injured diseased vessels.
Platelets are important in hemostasis; thrombocytopenic animals or
patients bleed extensively from injured vessels. Platelets are
also involved in the development of thrombi, particularly in the
arterial circulation, but to some extent in the venous circulation.
In addition, platelets contribute to the vessel wall changes associated with injury. Materials from platelets can increase permeability, or contribute to the repair that involves smooth muscle
cell migration and proliferation. In all of these interactions the
platelets adhere to surfaces, become adhesive and stick to each
other, discharge their granule contents, and synthesize products
from arachidonic acid freed from membrane phospholipids.

NORMAL VESSEL WALL

Although platelets do not adhere to normal endothelium, they
appear to be necessary for maintaining the integrity of the vascular
tree since hemorrhages occur throughout the microvascular bed in
humans and animals who are thrombocytopenic (1). At present, we
do not know what the platelets contribute to normal vessels to help
maintain the integrity of the vascular tree. Presumably in the
microcirculatory bed there are stresses in the vessel wall, perhaps
around vessel bifurcations, leading to disruption or changes in the

endothelium with the exposure of gaps which are plugged by platelets. If animals are made thrombocytopenic and platelets are then infused, platelets are found adhering to the vessel wall apparently at junction points between the endothelial cells (1). It has been proposed that the glycosaminoglycans that are associated with the surface of platelets may be involved in maintaining the integrity of the vessel lining (2).

When a small area of endothelium is removed, the surface is rapidly recovered by spreading of the endothelial cells. Some studies indicate that the presence of platelets, or a factor from platelets enhances this spreading (3). If platelets do make such a factor available, this could be one way in which platelets contribute to maintenance of the integrity of the vessel wall.

Stimulation or alteration of the endothelium leads initially to increased permeability. This is associated with increased accumulation of fluid and plasma proteins in the subendothelium, and can occur without any red cell accumulation (1,4). If, however, the gap between the endothelial cells in the microcirculation is sufficiently large, red cells can pass through and accumulate in the extravascular spaces. However, within large blood vessels with an intact subendothelium, removal of the endothelium does not lead to red cell accumulation in the wall.

ALTERED ENDOTHELIUM

As indicated in the previous section, platelets do not interact with normal endothelium. In addition, normal endothelium can bind thrombin and neutralize it, and take part in the activation of protein C and its anticoagulant action (5,6). Heparan sulfate associated with the surface of the endothelium is believed to be important in the ability of the endothelium to neutralize thrombin (5).

Endothelium can be altered in a variety of ways without desquamation. Exposure to some viruses can lead to transformation of the endothelium. Platelets can adhere to this altered endothelium (7). Recently, it has been observed that following exposure of rats to cigarette smoke, platelets adhere to the surface of the aorta, implying that the platelets are probably sticking to the endothelium as a consequence of some alteration produced by the effects of products from cigarette smoke (8).

Antibodies can interact with the endothelium, particularly in
the vessels of transplanted organs (9). Under these circumstances,
platelets adhere to the antibody-stimulated endothelial cells, possibly through interaction with immune complexes associated with the
surface of the endothelium. Leukocyte accumulation also occurs.
The leukocytes may adhere to the antigen-antibody complexes on the
surface of the endothelial cells or to the adherent platelets.
Leukocytes are known to adhere to the endothelium when the fifth
component of complement is activated, but this does not appear to
involve platelets (9).

DESQUAMATION OF ENDOTHELIUM

Breaks in the endothelial lining of vessels produce a variety
of responses depending upon where the break occurs and the degree
of injury to the vessel wall. For example, when the endothelium
is removed from large arteries, but the subendothelial layer is not
disrupted, red cells will not penetrate into the vessel wall, but
a layer of platelets forms on the surface of the exposed endothelium.
In areas where blood flow is disturbed, a mass of platelets accumulates on the adherent platelets. However, breaks in the endothelium associated with tearing or cutting of the vessel wall are
associated with extensive red cell loss which is only arrested if
the injury site is closed. The injury site can be closed by contraction of the vessel wall, the formation of a platelet-fibrin mass,
the formation of a blood clot with few platelets in it, or a combination of these (1,9). Arrest of bleeding from damaged normal vessels
is referred to as hemostasis and the platelet-fibrin mass that ar-
rests bleeding through the injury site is called a hemostastic plug.
When this reaction occurs on the inner lining of a blood vessel,
the mass that forms tends to protrude into the lumen and is usually
referred to as a thrombus. Damage to the vessel wall is usually
associated with acute contraction of the vessel at the injury site.
This contraction is believed to be important in hemostasis and may,
under some circumstances, be involved in thrombus formation in
diseased arteries. In all of these responses of the vessel wall to
injury there are common pathways. Thus, it is not surprising that
drugs that are effective inhibitors of thrombus formation may also
cause bleeding (1).

PLATELET ADHERENCE TO THE VESSEL WALL

When the endothelium is lost from the surface of normal blood vessels, platelets adhere to collagen fibers, microfibrils, and exposed basement membrane and spread on these subendothelial structures. Interaction of platelets with collagen leads to activation of the arachidonate pathway and release of the contents of some of the platelet granules (10). The role of plasma proteins in the interaction of platelets with subendothelial constituents is unclear. Albumin decreases the extent of platelet adhesion to collagen. Von Willebrand factor has also been found to be important in platelet adhesion to the subendothelium, but only in situations where there is a high shear rate (11), although von Willebrand factor may also facilitate the spreading of the platelets on the injured surface, probably on the microfibrils (12). At the shear rates that are found in large arteries, such as the aorta or the coronary arteries, it does not appear that von Willebrand factor has a major role in platelet adherence to exposed subendothelium.

If the endothelium is lost from the surface of a vessel without a break in the vessel wall, a thin layer of platelets forms on the exposed surface in areas where flow is not disturbed. The surface formed by the adherent platelets quickly becomes non-reactive to further platelet accumulation (9,10). In an area where flow is disturbed, however, flow can facilitate the formation of a platelet mass adherent to the vessel wall because of the accumulation of agents that stimulate platelets to become adhesive to each other and to the platelets adherent to the vessel wall (9). Factors contributing to flow probably include released ADP and serotonin from the amine storage granules, thromboxane A_2 formed from arachidonic acid, and trace amounts of thrombin that may be generated at such a site. Both thromboxane A_2 and thrombin can stimulate further platelet release and, therefore, further augment the aggregation process.

Injury to diseased or previously damaged vessels with a thickened intima leads to a somewhat different process. Under these conditions, the injured intimal smooth muscle cells appear to initiate coagulation, probably through tissue thromboplastin. This leads to the formation of thrombin and fibrin and the platelets do not appear to adhere directly to the damaged cells, but rather to the fibrin that forms on the surface of the injured cells. Platelet aggregates may form in association with the platelets that interact with the fibrin (13). This process may also occur in some forms of hemostasis where the injury leads to extensive damage to cells in the tissue and local activation of the coagulation mechanisms in addition to the normal interaction of the platelets with the exposed connective tissue. This type of response is dependent upon thrombin formation; thrombin induces platelet aggregation and the platelet release reaction through pathways that are independent of the release of ADP and the formation of thromboxane A_2 (10). Thus, the thrombin-induced reaction is difficult to inhibit with drugs that inhibit the cyclo-oxygenase of platelets. Thrombosis that is largely mediated by thrombin can be inhibited by oral anticoagulants or heparin (14).

DURATION OF ACTIVITY OF THE INJURY SITE

When the endothelium is removed from the surface of a large artery, the initial layer of platelets that forms becomes relatively non-reactive to the circulating platelets. This does not appear to be due to PGI_2, since inhibition ot PGI_2 formation after the injury does not cause increased platelet accumulation (15).

If the initial layer of platelets is prevented from forming on the vessel wall by infusion of PGI_2 or administration of di-pyridamole, platelet accumulation will occur to the usual extent if the therapy is stopped during the first 6 to 8 hours after injury. However, termination of the therapy after 8 hours does not result in extensive platelet accumulation on the surface (16). Thus, through mechanisms that are not yet understood, changes occur in the subendothelium that eventually make it non-reactive to significant further platelet accumulation (9,10). It has been shown that it is not due to PGI_2 formation because the adherence cannot be affected by inhibition of PGI_2 formation by the vessel wall. It seems unlikely that the accumulation of plasma proteins on the damaged wall results in it becoming non-reactive, since if platelets are kept from interacting with the damaged wall for periods up to 6 hours (during which time plasma proteins would have ample opportunity to coat the damaged surface) the usual amount of platelet accumulation occurs when the therapy is stopped.

It has also been observed that injury to diseased or previously injured vessels that results in the formation of platelet-fibrin

thrombi is also associated with loss of vessel wall reactivity (13).
It would appear that the mechanism that activates coagulation,
leading to localized thrombin formation, acts transiently. If
heparin is administered for a period of 6 to 8 hours to prevent
thrombus formation on a damaged artery, cessation of the heparin
therapy after 6 to 8 hours does not result in extensive thrombus
formation, whereas, if the therapy is stopped earlier, extensive
thrombus formation occurs (17).

TURNOVER OF THROMBI

Thrombi tend to form and undergo dissolution repeatedly at an
injury site during the period that the site remains active.
Platelets that are involved in thrombus formation probably escape
from the thrombus and return to the circulation (8). Experimental
evidence indicated that these platelets, although possibly degranulated and decreased in density, do not always have a shortened
survival.

Platelets may be freed from a thrombus in several ways. If
the release reaction has not been extensive, the platelets may
deaggregate. Alternatively, their adherence to fibrin may be
disrupted. Since platelet stimulation is necessary for the retraction of fibrin clots, it seems likely that platelet adherence to
fibrin is probably at least in part mediated through the same
mechanism as platelet interaction with fibrinogen. It follows that
if the platelet fibrinogen/fibrin receptors become unavailable, as
they do when platelets deaggregate, the platelets may no longer
remain adherent to the fibrin unless another mechanism comes into
play. If their attachment to fibrin is broken, blood flow may free
them from a thrombus even if little plasmin has formed to lyse the
fibrin. If the fibrin that stabilizes the thrombus is lysed, deag-
gregated platelets will be freed. Platelets that have adhered to
the vessel wall and then been detached may be rapidly cleared from
the circulation. Platelets may be freed from the wall by proteolytic enzymes, such as plasmin, or elastase from leukocytes that
disrupt the binding between platelets and the exposed connective
tissue at an injury site (19). If the platelets are freed in this
way, they may lose glycopeptides from their membrane glycoproteins,
with the result that their survival is shortened. Removal of the
glycopeptides from surface glycoproteins has been shown to cause
the rapid clearance of platelets from the circulation (20).

SPASM AND THROMBOSIS

When vessels are injured they contract (1). This contractile response is important in hemostasis since it can facilitate closure of a cut vessel. The degree of contraction of a vessel, however, probably represents a balance between factors that stimulate contraction and factors such as PGI_2 that cause relaxation. In animals that are thrombocytopenic, inhibition of PGI_2 formation by the vessel wall is associated with contraction of the puncture wounds in veins with arrest of bleeding. However, in circumstances where PGI_2 formation, which is probably stimulated at the puncture site, is not inhibited, the wound sites will stay open and in thrombocytopenic animals blood loss through them can be extensive (21). In coronary arteries, spasm may be important in causing myocardial ischemia and facilitating thrombosis. In a stenotic coronary artery, injury to the endothelium on the proximal side of the stenosis may be associated with spasm that promotes thrombus formation on the distal side of the stenosis (22,23). The nature of this relationship between spasm and thrombosis is unclear in the arterial system, but is well-established in hemostasis. It has been shown in the microcirculation that contraction of the vessels when they are injured is important in hemostasis, and if the contractile response is inhibited, the arrest of bleeding is less efficient. In these circumstances, aspirin in high doses that inhibits PGI_2 formation has been reported to shorten the bleeding time (24).

PLATELETS, ENDOTHELIUM AND THE VESSEL WALL

When platelets interact with an injured vessel wall, they release factors that promote smooth muscle cell migration and proliferation (9,10). This will be covered in more detail by Russell Ross. This proliferative response of the smooth muscle cells can also be modulated by the endothelium, which can form agents that inhibit smooth muscle cell proliferation, although there are some circumstances in which the endothelium can be stimulated to form factors that promote smooth muscle cell proliferation (25).

REFERENCES

1.   J.F. Mustard and M.A. Packham, The reaction of the blood in injury, in: "Inflammation, Immunity and Hypersensitivity," Second edition, H.Z. Movat, ed., Harper & Row, New York,

pp. 557–664 (1978).

2. C.S. Kitchens and L. Weiss, Ultrastructural changes of endo-
    thelium associated with thrombocytopenia, Blood, 46:567–578
    (1975).

3. A.I. Gotlieb and M.K.K. Wong, The effect of platelet poor
    plasma serum on endothelial cell spreading, Artery, 9:59–68
    (1981).

4. G. Majno and I. Joris, Endothelium 1977: a review, in: "The
    Thrombotic Process in Atherogenesis," A.B. Chandler, K.
    Eurenius, G.C. McMillan, C.B. Nelson, C.J. Schwartz, S.
    Wessler, eds., Adv. Exp. Med. Biol., 104:169–225 (1978).

5. R. Ross, Atherosclerosis: a problem of the biology of arterial
    wall cells and their interactions with blood components,
    Arteriosclerosis, 1:293–311 (1981).

6. P.C. Comp, R.M. Jacocks, G.L. Ferrell, and C.T. Esmon, Activa-
    tion of protein C in vivo, J. Clin. Invest., 70:127–134
    (1982).

7. K.D. Curwen, H.-Y. Kim, M. Vazquez, R.I. Handin, and M.A.
    Gimbrone Jr., Platelet adhesion to cultured vascular endo-
    thelial cells. A quantitative monolayer adhesion assay,
    J. Lab. Clin. Med., 100:425–436 (1982).

8. R.M. Pittilo, I.J. Mackie, P.M. Rowles, S.J. Machin, and N.
    Woolf, Effects of cigarette smoking on the ultrastructure
    of rat thoracic aorta and its ability to produce prosta-
    cyclin, Thromb. Haemost., 48:173–176 (1982).

9. J.F. Mustard, M.A. Packham, and R.L. Kinlough-Rathbone,
    Platelets, atherosclerosis, and clinical complications,
    in: "Vascular Injury and Atherosclerosis," S. Moore, ed.,
    Marcel Dekker, New York, pp. 79–110 (1981).

10. J.F. Mustard, M.A. Packham, and R.L. Kinlough-Rathbone,
    Mechanisms in thrombosis, in: "Haemostasis and Thrombosis,"
    A.L. Bloom and D.P. Thomas, eds., Churchill Livingstone,
    London, pp. 503–526 (1982).

11. H.J. Weiss, V.T. Turitto, and H.R. Baumgartner, Effect of
    shear rate on platelet interaction with subendothelium
    in citrated and native blood. 1. Shear rate-dependent
    decrease of adhesion in von Willebrand's disease and
    Bernard-Soulier syndrome, J. Lab. Clin. Med., 92:750–764
    (1978).

12. Y.J. Legrand, F. Fauvel, N. Gutman, J.P. Muh, G. Tobelem, H.
    Souchon, A. Karniguian, and J.P. Caen, Microfibrils (mf)
    platelet interaction: requirement of von Willebrand factor,

Thromb. Res., 19:737-739 (1980).

13. H.M. Groves, R.L. Kinlough-Rathbone, M. Richardson, L. Jørgensen, S. Moore, and J.F. Mustard, Thrombin generation and fibrin formation following injury to rabbit neointima: studies of vessel wall reactivity and platelet survival, Lab. Invest., 46:605-612 (1982).

14. M.A. Packham and J.F. Mustard, Pharmacology of platelet-affecting drugs, Circulation, 62:(Suppl. V) V-26-V-41 (1980).

15. E. Dejana, J.-P. Cazenave, H.M. Groves, R.L. Kinlough-Rathbone, M. Richardson, M.A. Packham, and J.F. Mustard, The effect of aspirin inhibition of PGI$_2$ production on platelet adherence to normal and damaged rabbit aortae, Thromb. Res., 17:453-464 (1980).

16. H.P. Groves, R.L. Kinlough-Rathbone, and J.F. Mustard, Effect on vessel wall reactivity of inhibiting platelet accumulation with dipyridamole, Circulation, 66:11-53 (1982).

17. D.G. Piepgras, T.M. Sundt Jr., and P. Didisheim, Effect of anticoagulants and inhibitors of platelet aggregation on thrombotic occlusion of endarterectomized cat carotid arteries, Stroke, 7:248-254 (1976).

18. P.D. Winocour, M. Cattaneo, D. Somers, M. Richardson, R.L. Kinlough-Rathbone, and J.F. Mustard, Platelet survival and thrombosis, Arteriosclerosis, 2:458-466 (1982).

19. P.D. Winocour, R.L. Kinlough-Rathbone, M. Richardson, and J.F. Mustard, Reversal of shortened platelet survival in rats by the antifibrinolytic agent, epsilon aminocaproic acid, J. Clin. Invest., 71:159-164 (1983).

20. J.P. Greenberg, M.A. Packham, M.A. Guccione, M.L. Rand, H.-J. Reimers, and J.F. Mustard, Survival of rabbit platelets treated in vitro with chymotrypsin, plasmin, trypsin, or neuraminidase, Blood, 53:916-927 (1979).

21. M.A. Blajchman, A.F. Senyi, J. Hirsh, Y. Surya, and M. Buchanan, Shortening of the bleeding time in rabbits by hydrocortisone caused by inhibition of prostacyclin generation by the vessel wall, J. Clin. Invest., 63:1026-1035 (1979).

22. J.F. Mustard, R.L. Kinlough-Rathbone, and M.A. Packham, The vessel wall in thrombosis, in: "Hemostasis and Thrombosis. Basic Principles and Clinical Practice," R.W. Colman, J. Hirsh, V.J. Marder, and E.W. Salzman, eds., J.B. Lippincott, Philadelphia, pp. 703-715 (1982).

23. J.F. Mustard, R.L. Kinlough-Rathbone, and M.A. Packham,

    Atherogenesis and its relation to stroke, <u>Clin. Neurosurg.</u>,
    29:417-436 (1982).
24. J. O'Grady and S. Moncada, Aspirin: a paradoxical effect on
    bleeding time, <u>Lancet</u>, 2:780 (1978).
25. J.J. Castellot Jr., M.L. Addonizio, R. Rosenberg, and M.J.
    Karnovsky, Cultured endothelial cells produce a heparin-
    like inhibitor of smooth muscle cell growth, <u>J. Cell. Biol.</u>,
    90:372-379 (1981).

ROLES OF THROMBOSPONDIN AND FIBRONECTIN

IN VASCULAR INJURY

Deane F. Mosher and Paula J. McKeown-Longo

Department of Medicine
University of Wisconsin
Madison, Wisconsin
U.S.A.

Thrombospondin is a major platelet α-granule glycoprotein that is secreted and then bound to platelet membranes when platelets aggregate in response to thrombin (1-5). Studies by Lawler and others (6,9) indicate that thrombospondin is 420-450 kilodaltons (kd), has a pI of 4.7, and is composed of three large disulfide-linked subunits. It interacts avidly with heparin (7,10).

During the process of aggregation, thrombin-stimulated platelets develop a membrane-bound activity which allows the platelets to agglutinate fixed trypsinized erythrocytes (11-13). This activity originates from α-granules and appears to play an important role in mediating platelet aggregation by binding to a specific receptor on other platelets (14,15). Jaffe et al. (16) recently found that purified human platelet thrombospondin has similar activity (i.e., it agglutinates fixed trypsinized sheep erythrocytes and influences the agglutination of erythrocytes by thrombin-treated human platelets). Therefore, it is likely that thrombospondin is the endogenous "lectin" of platelets (16). Thrombospondin preadsorbed to ex vivo polymeric shunts causes rapid binding and activation of platelets on the polymeric surface (17). The use of chemical cross-linkers (18) has implicated thrombospondin in the attachment of platelets to fibronectin and collagen-coated substrata. Binding studies using purified proteins have indicated that thrombospondin specifically binds fibrinogen and fibronectin (19), type V collagen (20), and histidine-rich glycoprotein (21). Although much more

needs to be learned about the biologic importance of these interactions, these results suggest that thrombospondin may play an
important role in hemostasis.

Cultured endothelial cells (9,22), fibroblasts (23,24), and
smooth muscle cells (23) synthesize and secrete glycoprotein which
is a disulfide-bonded multimer of 160-190 subunits, is not collagenous, and has recently been identified as thrombospondin.
Thrombospondin can be detected by immunofluorescence in the extracellular fibrillar matrix of these cells (23,24).

Fibronectin is a large glycoprotein which is found in most
tissues and body fluids. It is a worm-like molecule with considerable segmental flexibility (25-29). There have been extensive
studies of fibronectin synthesized by cultured cells and fibronectin
isolated from plasma (30-32). Both types of fibronectin are 400-500
kd disulfide-bonded dimers of similar 200-500 kd subunits. Although
both cell culture and plasma fibronectin have specific binding
sites for collagen, fibrin, hyaluronic acid, heparin, staphylococci,
and actin, differences have been detected for some structural (33),
immunological (34), and biological (35) properties.

Cellular fibronectin exists as an insoluble connective tissue
protein found on cell surfaces and in the extracellular matrix.
Fibronectin probably functions as an adhesive protein for cell
attachment and tissue organization (30-32). Fibronectin synthesized
in cell culture is found both in the culture medium and deposited
in the cell layer where it forms disulfide-bonded multimers. Cell
surface and extracellular fibronectin is found in close association
with collagen, heparan sulfate proteoglycans, hyaluronic acid, and
other proteins making up a detergent-insoluble extracellular
matrix (36-41).

Fibronectin is present in a soluble form at levels of 300
µg/ml in human plasma. Most circulating fibronectin is probably
secreted by hepatocytes (42). The functions of plasma fibronectin
are uncertain. It becomes incorporated into the clot during
coagulation, and may serve a role in wound healing as an attachment
site for cells invading the clot (43). In vivo experiments have
indicated that plasma fibronectin can be incorporated into tissue
extracellular matrices (44). Thus, plasma fibronectin may serve
as a reservoir for tissue fibronectin. This theory is supported
by in vitro observation that serum fibronectin from serum-
supplemented cell culture medium is incorporated into the cell

Table I.   Accumulation of Thrombospondin (TSP) and Fibronectin (FN)
           over one to three Days in Medium of Cultured Adherent Cells

Cells	TSP $\mu g/10^6$			FN $\mu g/10^6$		
	$\mu g/ml$	Cells	Ref.	$\mu g/ml$	Cells	Ref.
Human umbilical vein endothelial cells	2.2-4.4	21-49	9,24	5.1	ND	51
Human skin fibroblasts	2.2	2.5-15.7	23,24	2.8	ND	52
Human aortic smooth muscle cells	ND	7.1	23			
Bovine aorta endothelial cells	3.6	1.0	23[a]			

[a]E.A. Jaffe and D.F. Mosher, unpublished experiment.
ND - Not Determined

layer (45-47). Although it is unusual for plasma or serum proteins
to become incorporated into connective tissues, such behavior is
nót unique to fibronectin and thrombospondin. Amyloid P component
(48) (plasma concentration of 30 mg/l), von Willebrand factor (49)
(plasma concentration of 10 mg/l), and an adhesive protein called
vitronectin (50) (plasma concentration of 200 mg/l) have been
localized to connective tissues by immunomicroscopy and/or extraction.

     Similar amounts of thrombospondin and fibronectin accumulate in
serum-containing growth medium as a result of synthesis and secretion by cultured cells, including endothelial cells (Table I).
In addition, the serum-containing medium itself contains thrombospondin and fibronectin. Plasma thrombospondin concentration is
low (53) (also Table II), and most of the thrombospondin in serum
is thrombospondin that is released from platelets during clotting.
The opposite is true for fibronectin. Most of the fibronectin in
whole blood is in plasma, and a fraction of this becomes cross-
linked to fibrin and lost during clotting (30,54). As a result,

Table II.   Thrombospondin Concentration in Human Serum and Plasma

Sample	Concentration (μg/ml)	% of Serum Concentration
"Thrombotect[TM]" Plasma	0.34	1.0
Citrated Plasma	1.2	3.4
Serum	34.3	100.0

Thrombospondin concentration was measured by ELISA (24).   The "Thrombotect[TM]" system minimized release of platelet α-granule proteins during the preparation of plasma.   The concentration of thrombospondin standard was calculated using a published extinction co-efficient (7).

the measured concentration of fibronectin in 10% fetal calf serum, 3-4 μg/ml (45), is very similar to the concentration that can be calculated for thrombospondin.

There were remarkable differences between the metabolism of ^{125}I-fibronectin and ^{125}I-thrombospondin by RuBa human endothelial cells (Table III).   Fibronectin became deposited in the cell matrix in two pools, one which was soluble in 1% deoxycholade and a second which was insoluble in 1% deoxycholate but soluble in 2% sodium dodecyl sulfate.   Very little fibronectin was degraded. In contrast, ^{125}I-thrombospondin was largely bound in the pool which was soluble in 1% deoxycholate, and thrombospondin from the medium was rapidly degraded.

This experiment was done with confluent RuBa cells (55) grown in 28-cm^2 dishes (8 X 10^5 cells/dish).   Cell layers were washed and incubated with 2 ml of F12 medium containing 10% fibronectin-depleted human serum and the amounts of ^{125}I-labeled protein indicated in parentheses.   At the indicated times, cultures were split into media and washed cell layers.   Labeled protein in the cell layer was extracted with 1% deoxycholate (DOC) and then with 2% sodium dodecyl sulfate.   Labeled protein in the medium was precipitated with 10% trichloroacetic acid (TCA) to determine how much was degraded.

Table III. Metabolism of ^{125}I-Thrombospondin (^{125}I-TSP) and ^{125}I-Fibronectin (^{125}I-FN) Added to Media of Confluent RuBa Human Endothelial Cells

Labeled Protein	Hours	% Label		
		DOC-Soluble Pool	DOC-Soluble Pool	TCA-Soluble
^{125}I-TSP	3	5.8	< 0.5	5.3
(4 µg/ml)	6	4.9	< 0.5	11.4
	12	6.9	< 0.5	22.3
	24	2.7	< 0.5	34.3
^{125}I-FN	3	5.7	3.0	< 0.2
(2 µg/ml)	6	4.9	4.0	< 0.2
	12	3.5	5.9	0.4
	24	2.6	5.8	0.6

The differences in metabolism between fibronectin and thrombospondin indicate that these two proteins play different roles, although both are found in the extracellular fibrillar matrix and seem to have the property of mediating cell adhesion. In a wounded area, both thrombospondin and fibronectin are likely to be present in high concentration as soluble proteins. In wound areas, there may be interactions among thrombospondin, fibronectin, and cell surfaces of activated cells (e.g., cells stimulated to divide by exposure to platelet growth factor and to migrate in response to platelet factor 4 and β-thromboglobulin) (56) which result in the insolubilization of thrombospondin and fibronectin and the orderly elaboration of a scaffolding, which can in turn support deposition of collagen and proteoglycans. Interestingly, thrombospondin has been shown to bind to fibronectin in a micro ELISA (57), and to become associated with fibronectin when platelets adhere to fibronectin-coated collagen fibrils (58).

REFERENCES

1.   I. Hagen, Effects of thrombin on washed, human platelets: Changes in subcellular fractions, Biochem. Biophys. Acta, 392:242–254 (1975).

2.   I. Hagen, T. Olsen, and N.O. Solum, Studies on subcellular
     fractions of human platelets by the lactoperoxidase iodina-
     tion technique, Biochem. Biophys. Acta, 455:214-225 (1976).

3.   J.W. Lawler, F.C. Chao, and P.-H. Fang, Observation of a high
     molecular weight platelet protein released by thrombin,
     Thrombos. Haemostas., 37:355-357 (1977).

4.   J.N. George, R.M. Lyons, and R.K. Morgan, Membrane changes
     associated with platelet activation. Exposure of actin
     on the platelet surface after thrombin-induced secretion,
     J. Clin. Invest., 66:1-9 (1980).

5.   D.R. Phillips, L.K. Jennings, and H.R. Prasanna, Ca^{2+}-mediated
     association of glycoprotein G (thrombin-sensitive protein,
     thrombospondin) with human platelets, J. Biol. Chem.,
     255:11629-11632 (1980).

6.   J.W. Lawler, H.S. Slayter, and J.E. Coligan, Isolation and
     characterization of a high molecular weight glycoprotein
     from human blood particles, J. Biol. Chem., 253:8609-8616
     (1978).

7.   S.S. Margossian, J.W. Lawler, and H.S. Slayter, Physical
     characterization of platelet thrombospondin, J. Biol. Chem.,
     256:7495-7500 (1981).

8.   J. Lawler, F.C. Chao, and C.M. Cohen, Evidence for calcium-
     sensitive structure in platelet thrombospondin: isolation
     and partial characterization of thrombospondin in the
     presence of calcium, J. Biol. Chem., 257:12257-12265 (1982).

9.   D.F. Mosher, M.J. Doyle, and E.A. Jaffe, Synthesis and secre-
     tion of thrombospondin by cultured human endothelial cells,
     J. Cell Biol., 93:343-348 (1982).

10. J.W. Lawler and H.S. Slayter, The release of heparin-binding
     peptides from platelet thrombospondin by proteolytic action
     of thrombin, plasmin and trypsin, Thromb. Res., 22:267-279
     (1981).

11. T.K. Gartner, D.C. Williams, and D.R. Phillips, Platelet plasma
     membrane lectin activity, Biophys. Res. Commun., 79:592-599
     (1977).

12. T.K. Gartner, D.C. Williams, F.C. Minion, and D.R. Phillips,
     Thrombin-induced platelet aggregation is mediated by a
     platelet plasma membrane-bound lectin, Science, 200:1281-
     1283 (1978).

13. T.K. Gartner, D.R. Phillips, and D.C. Williams, Expression of
     thrombin-enhanced platelet lectin activity is controlled
     by secretion, FEBS Lett., 113:196-200 (1980).

14. T.K. Gartner, J.M. Gerrard, J.G. White, and D.C. Williams, Fibrinogen is the receptor for the endogenous lectin of human platelets, Nature, 289:688-700 (1981).

15. T.K. Gartner, J.M. Gerrard, J.G. White, and D.C. Williams, The endogenous lectin of human platelets is an α-granule component, Blood, 58:153-157 (1981).

16. E.A. Jaffe, L.L.K. Leung, R.L. Nachman, R.I. Levin, and D.F. Mosher, Thrombospondin is the endogenous lectin of human platelets, Nature, 295:246-248 (1981).

17. B.R. Young, M.J. Doyle, W.E. Collins, L.K. Lambrecht, C.A. Jordan, R.M. Albrecht, D.F. Mosher, and S.L. Cooper, Effects of thrombospondin and other platelet α-granule proteins on artificial surface-induced thrombosis, Trans. Am. Soc. Art. Intern. Organs, 28:498-503 (1982).

18. J. Lehav, M.A. Schwartz, and R.O. Hynes, Analysis of platelet adhesion with a radioactive chemical cross-linking reagent: interaction of thrombospondin with fibronectin and collagen, Cell, 31:253-262 (1982).

19. L.L.K. Leung and R.L. Nachman, Complex formation of platelet thrombospondin with fibrinogen, J. Clin. Invest., 70:542-549 (1982).

20. S.M. Mumby, G.J. Raugi, and P. Bornstein, Interactions of thrombospondin with extracellular matrix proteins: selective binding to type V collagen, J. Cell Biol. (in press) (1984).

21. L.L.K. Leung, R.L. Nachman, and P.C. Harpel, Complex formation of platelet thrombospondin with histidine-rich glyco-protein, J. Clin. Invest., (in press) (1984).

22. J. McPherson, H. Sage, and P. Bornstein, Isolation and characterization of a glycoprotein secreted by aortic endothelial cells in culture. Apparent identity with platelet thrombospondin, J. Biol. Chem., 256:11330-11336 (1981).

23. G.J. Raugi, S.M. Mumby, D. Abbott-Brown, and P. Bornstein, Thrombospondin: synthesis and secretion by cells in culture, J. Cell Biol., 95:351-354 (1982).

24. E.A. Jaffe, J.T. Ruggiero, L.L.K. Leung, M.J. Doyle, P.J. McKeown-Longo, and D.F. Mosher, Cultured human fibroblasts synthesize and secrete thrombospondin and incorporate it into extracellular matrix, Proc. Natl. Acad. Sci. USA, (1983).

25. M.W. Mosesson, A.B. Chen, and R.M. Huseby, The cold-insoluble

globulin of human plasma: studies of its essential
structural features, Biochem. Biophys. Acta, 386:509-524
(1975).

26. E.C. Williams, P.A. Janmey, J.D. Ferry, and D.F. Mosher,
Conformational states of fibronectin: effects of pH,
ionic strength, and collagen binding, J. Biol. Chem.,
257:14973-14978 (1982).

27. S.S. Alexander, Jr., G. Colonna, and H. Edelhoch, The structure
and stability of human plasma cold-insoluble globulin,
J. Biol. Chem., 254:1501-1505 (1979).

28. J. Engel, E. Odermatt, A. Engel, J. Madri, H. Furthmayr, H.
Rohde, and R. Timpl, Shapes, domain organizations and
flexibility of laminin and fibronectin, two multifunctional
proteins of the extracellular matrix, J. Mol. Biol., 150:
97-120 (1981).

29. H.P. Erickson, N. Carrell, and J. McDonagh, Fibronectin molecule
visualized in electron microscopy: a long, thin, flexible
strand, J. Cell Biol., 91:673-678 (1981).

30. D.F. Mosher, Fibronectin, Prog. Hemostas. Thrombosis, 5:111-
151 (1980).

31. R.O. Hynes and K.M. Yamada, Fibronectins: Multifunctional
modular glycoproteins, J. Cell Biol., 95:369-377 (1982).

32. E. Ruoslahti, E. Engvall, and E.G. Hayman, Fibronectin:
current concepts of its structure and functions, Collagen
Res., 1:95-128 (1981).

33. M. Hayashi and K.M. Yamada, Differences in domain structures
between plasma and cellular fibronectins, J. Biol. Chem.,
256:11292-11300 (1981).

34. B.T. Atherton and R.O. Hynes, A difference between plasma and
cellular fibronectins located with monoclonal antibodies,
Cell, 25:133-141 (1981).

35. K.M. Yamada and D. Kennedy, Fibroblast cellular and plasma
fibronectins are similar but not identical, J. Cell Biol.,
80:492-498 (1979).

36. L.B. Chen, A. Murray, R.A. Segal, A. Bushnell, and M.L. Walsh,
Studies on intracellular LETS glycoprotein matrices,
Cell, 14:377-391 (1978).

37. K. Hedman, M. Kurkinen, K. Alitalo, A. Vaheri, S. Johansson,
and M. Höök, Isolation of the pericellular matrix of
human fibroblasts cultures, J. Cell Biol., 81:83-91 (1979).

38. W.G. Carter and S-I. Hakomori, A new cell surface, detergent-
insoluble glycoprotein matrix of human and hamster fibroblasts. The role of disulfide bonds in stabilization of

the matrix, J. Biol. Chem., 256:6953-6960 (1981).

39. W.G. Carter, The cooperative role of the transformation-sensitive glycoproteins, GP 140 and fibronectin, in cell attachment and spreading, J. Biol. Chem., 257:3249-3257 (1982).

40. E.G. Hayman, Å. Oldberg, G.R. Martin, and E. Ruoslahti, Co-distribution of heparan sulfate proteoglycan, laminin, and fibronectin in the extracellular matrix of normal rat kidney cells and their coordinate absence in transformed cells, J. Cell Biol., 94:28-35 (1982).

41. K. Hedman, S. Johansson, T. Vartio, L. Kjellén, A. Vaheri, and M. Höök, Structure of the pericellular matrix: association of heparan and chondroitin sulfates with fibronectin-procollagen fibres, Cell, 28:663-671 (1982).

42. M.R. Owens and C.D. Cimino, Synthesis of fibronectin by the isolated perfused rat liver, Blood, 59:1305-1309 (1982).

43. F. Grinnell, M. Feld, and D. Minter, Fibroblast adhesion to fibrinogen and fibrin substrata: requirement for cold insoluble globulin (plasma fibronectin), Cell, 19:517-525 (1980).

44. E. Oh, M. Pierschbacher, and E. Ruoslahti, Deposition of plasma fibronectin in tissues, Proc. Natl. Acad. Sci. USA, 78:3218-3221 (1981).

45. E.G. Hayman and E. Ruoslahti, Distribution of fetal bovine serum fibronectin and endogenous rat cell fibronectin in extracellular matrix, J. Cell Biol., 83:10-14 (1979).

46. S.I. Rennard, M.L. Wind, A.T. Hewitt, and H.K. Kleinman, Effect of collagen and cell shape on binding of fibronectin to cells, Arch. Biochem. Biophys., 206:205-212 (1981).

47. P.J. McKeown-Longo and D.F. Mosher, The binding of plasma fibronectin to cell layers of human skin fibroblasts, J. Cell Biol., 97:446-472 (1983).

48. R.F. Dyck, C.M. Lockwood, M. Kershaw, N. McHugh, V. Duance, M.L. Baltz, and B. Pepys, Amyloid P component is a constituent of normal human glomerual basement membrane, J. Exp. Med., 152:1162-1174 (1981).

49. J.H. Rang, I.I. Sussman, R.E. Gordon, S.V. Chu, and V. Solomon, Localization of factor VIII-related antigen in human vascular subendothelium, Blood, 55:752-756 (1980).

50. E.G. Hayman, M.D. Pierschbacher, Y. Öhgren, and E. Ruoslahti, Serum spreading factor (Vitronectin) is present at the cell surface and in tissues, Proc. Natl. Acad. Sci. USA, 80:4003-4007 (1983).

51. E.A. Jaffe and D.F. Mosher, Synthesis of fibronectin by
     cultured human endothelial cells, J. Exp. Med., 147:1779–
     1791 (1978).

52. D.F. Mosher, O. Saksela, J. Keski-Oja, and A. Vaheri, Distribu-
     tion of a major surface-associated glycoprotein, fibronectin,
     in cultures of adherent cells, J. Supramolec. Struct.,
     6:551–557 (1977).

53. S.D. Saglio and H.S. Slayter, Use of a radioimmunoassay to
     quantify thrombospondin, Blood, 59:162–166 (1982).

54. D.F. Mosher, Action of fibrin-stabilizing factor on cold
     insoluble globulin and α_2-macroglobulin in clotting plasma,
     J. Biol. Chem., 251:1639–1645 (1976).

55. C.A. Reznikoff and R. DeMars, In vitro chemical mutagenesis
     and viral transformation of a human endothelial cell strain,
     Cancer Res., 41:1114–1126 (1981).

56. R.M. Senior, G.L. Griffin, J.S. Huang, D.A. Walz, and T.F.
     Deuel, Chemotactic activity of platelet alpha granule
     proteins for fibroblasts, J. Cell Biol., 96:382–385 (1983).

57. L.L. Lueng and R.L. Nachman, Complex formation of platelet
     thrombospondin with fibrinogen, J. Clin. Invest., 70:542–
     549 (1982).

58. J. Lehav, M.A. Schwartz, and R.O. Hynes, Analysis of platelet
     adhesion with a radioactive chemical crosslinking reagent:
     Interaction of thrombospondin with fibronectin and collagen,
     Cell, 31:253–262 (1982).

THE INTERACTION OF FIBRONECTIN, ENDOTHELIUM,

AND SUBENDOTHELIUM

Franco Piovella

Istituto di Clinica Medica I
Università di Pavia
Pavia, Italy

Fibronectin has been localized in a wide variety of tissue and cell types and can be considered as an ubiquitous protein (56). It can be extracted from blood, amniotic fluid, cerebrospinal fluid, seminal blood, fibroblasts, endothelial cells, myoblasts, lens epithelial cells, and platelets (1). Furthermore, fibronectin can be isolated from a wide range of basement membranes.

Two major forms of fibronectin have been described: plasma fibronectin and cellular fibronectin (34,40,52,57). Although both fibronectins have similar molecular weight (200,000 - 250,000), share immunological reactivity and biological activities, there are some differences between them, the most relevant being that cellular fibronectin is an insoluble multimer, whereas, in physiological conditions, plasma fibronectin is a soluble dimer. Important structural differences have also been identified utilizing monoclonal antibodies (2).

Many cell types in culture synthesize fibronectin, which is thought to be secreted either directly onto the cell surface or into the culture media (56). To date, it is not yet clear how secreted fibronectin becomes organized into its cellular form, which constitutes the fibrillar, extracellular matrix found in cultured fibroblasts, endothelial cells, smooth muscle cells, epithelial cells, astroglial cells, hepatocytes, and chondrocytes (16). Once fibronectin is placed on the cellular surface or in

the basement membrane, it can be lost by sloughing from cells or
by protease degradation (56). Cell surface fibronectin may be the
precursor of plasma fibronectin, and its release may be in response
to low extracellular levels. Fibroblasts, vascular endothelial
cells, hepatocytes, and the reticulo-endothelial system or macrophage
system have been at different times considered as the "major" site
of synthesis of plasma fibronectin (24,31,37,48,56). To date,
the "major" site of synthesis of plasma fibronectin still remains
undefined, although there is little doubt that one or all of these
cellular types contribute to maintaining the blood levels of this
protein.

Fibronectin is involved in a wide variety of cellular activities.
In both soluble or cellular form, it functions as an adhesive
protein, binding directly to collagen and binding cells to other
cells or to the substrata on which they are grown (19,39,57). It
also plays a role in the transformation process, i.e., there is
an inverse relationship between the expression of fibronectin and
cells transformation (22,51). Fibronectin is also a major opsonic
protein in plasma (47).

Roles for fibronectin in hemostasis have been recently envisaged. An attempt will be made here to define the relationships
existing between fibronectin, hemostasis, and thrombosis. The
functions of fibronectin will be described on the basis of its
location, i.e., intracellular, extracellular or soluble, considering
endothelium as a pivotal element.

The vascular endothelium constitutes a specialized "organ" with
unique physical and metabolic properties. It exists as a monolayer
of cells which constitutes the first interface between blood and
tissue. It is few fractions of a micron thick and several hundreds
of square meters wide. Its organization as a broad ultrathin monolayer amplifies the effectiveness and the rapidity of its metabolic
activities, but on the other hand it makes it very sensible to the
minimal injury. Studies of the functions of endothelial cells have
expanded rapidly in the past few years and several important
conceptual advances have been made. Our knowledge of the various
functions of endothelial cells has greatly increased, since it has
become possible to culture these cells (12,26). Their metabolic
activities have since been studied in cultures from human umbilical
cord vein and from aortas of pigs, cows, rabbits, and rats (6,7,21,
50).

It is well established that endothelial cells can either promote hemostasis or exert antithrombotic activities. Among the most relevant features which have been attributed to endothelium on the pro-hemostatic side, the synthesis of factor VIII-related antigen (VIIIR:Ag) and the synthesis of several components of the basement membrane are of particular relevance to this report.

Human endothelial cells can synthesize and release a great amount of soluble fibronectin (4,30,48). From experiments in vitro it has been calculated that the vascular endothelium as a whole can produce up to 2 g of fibronectin a day (48). Fibronectin has been considered as one of the major proteins synthesized and secreted by endothelial cells in culture (24). It constitutes about the 15% of the protein released by these cells in the culture media. As a comparison, we shall consider that VIIIR:Ag represents less than 1% of the total protein secreted by endothelial cells. This endothelial-produced fibronectin possesses subunits having the same molecular weight as those found in human plasma. Endothelial cells in culture also synthesize other components of the basement membrane and assemble it in an extracellular matrix. Indirect immunofluorescent studies showed that fibronectin is the major non-collagenous component of this matrix (4).

A very interesting characteristic of fibronectin is its existence both as an insoluble extracellular matrix protein and as a soluble protein in plasma. Endothelial cell cultures similarly contain insoluble fibronectin in the cell layer and soluble fibronectin in the medium.

Soluble fibronectin secreted by human endothelial cells contributes to maintaining the plasma levels of this protein. Circulating fibronectin plays a regulatory role in the blood coagulation process. It binds covalently to fibrin through the action of factor XIIIa (33,34), and about 3 to 4% of the normal clot consists of fibronectin. This accounts for the lower concentration of this protein usually found in serum. The distribution of fibronectin strands has been studied by immunofluorescence in freshly clotted blood. There are indications that fibrin strands formed in vivo are probably a fibrin-fibronectin copolymer which interacts with platelets (18). Fibronectin is thought to provide substantial mechanical stability and might be necessary for clot retraction. Fibronectin can also be cross-linked to itself by

factor XIIIa (34). It has also been suggested that factor XIIIa-
induced fibronectin polymerization may have an in vivo importance
in the formation of the fibrillar form of this protein. Clotting
of factor XIIIa-deficient plasma allows all fibronectin to remain
in serum. This phenomenon could be related to the defective wound
healing (see below) present in patients with factor XIIIa deficiency
(40).

Soluble fibronectin has been found in platelets which release
it following stimulation (14,58). After activation, specific
receptors appear on the platelet surface. These receptors seem to
be important in promoting platelet spreading (27). It has been
proposed that fibronectin may be the collagen receptor on platelet
(3). Nevertheless, pre-treatment of platelets with anti-fibronectin
antibodies shows little inhibition of adhesion of platelet to
collagen (49). More recently it has been demonstrated that
fibronectin is localized in α-granules from which it is released
following stimulation (13,45). In vitro aggregation of normal
platelets by collagen occurs without addition of exogenous source
of fibronectin. Nevertheless, it has been demonstrated that normal
plasma fibronectin restores to deficient plasma its ability to
support platelet aggregation by collagen, thereby implying a neces-
sity for fibronectin in normal platelet-collagen interactions (58).
It has been suggested that any fibronectin on the platelet surface,
whether it is adsorbed plasma fibronectin or a true peripheral
component of the platelet surface, will contribute to the adhesion
of platelet to collagen. Briefly, the occurrence in platelet of
two fibronectin pools is confirmed: an intracellular pool, localized
in α-granules which is not available nor exposed on platelet membrane
prior to activation, and a surface pool constituted by intracellular
fibronectin expressed by the platelet after binding to collagen
fibers, plus fibronectin recruited from exogenous sources (plasma)
by the spread platelet that brings the surface fibronectin to a
higher level than that reached by simple expression of endogenous
fibronectin (29). To date, it is not possible to postulate different roles for these two pools. However, both the membrane
expression and the binding of plasma fibronectin after surface
activation of the platelet confirm the possible participation of
this protein in blood coagulation associated with platelet membrane
reactions (29).

In addition of its function in blood coagulation, fibronectin

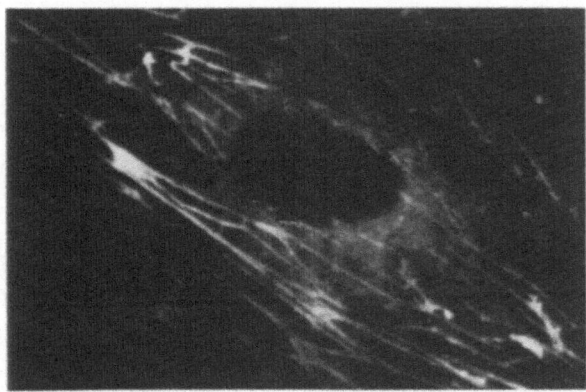

Fig. 1. Distribution of cell surface
fibronectin. Indirect immunofluorescence with rabbit anti-
fibronectin antiserum and FITC-
conjugated goat antirabbit
antiserum. Endothelial cell.
800 X.

in its soluble form is a major opsonin in plasma. All of the
specific binding domains of fibronectin are important in this
function. Fibronectin specifically promotes the binding of fibrin
to monocytes (27, thus clearing the fibrin from the circulation,
i.e., it mediates the clearance of soluble fibrin that escapes
the area of clotting. Plasma fibronectin is decreased after major
surgery or major trauma, and in severely ill patients with disseminated intravascular coagulation. In these circumstances, the
plasma concentration of fibronectin has been reported to be
decreased because of increased utilization rather than decreased
synthesis (35). Restoration of fibronectin concentration by
giving cryoprecipitate has a beneficial effect on the course of
the disease.

Insoluble fibronectin is distributed in three different
compartments: intracellularly, at the cell surface, and in the
extracellular matrix (10,11,20,46).

The intracellular pool of fibronectin represents synthesized
fibronectin about to be secreted, and it does not seem to be taken
up by endocytosis from the culture medium (which contains fetal
calf serum, and therefore, bovine fibronectin), because endothelial

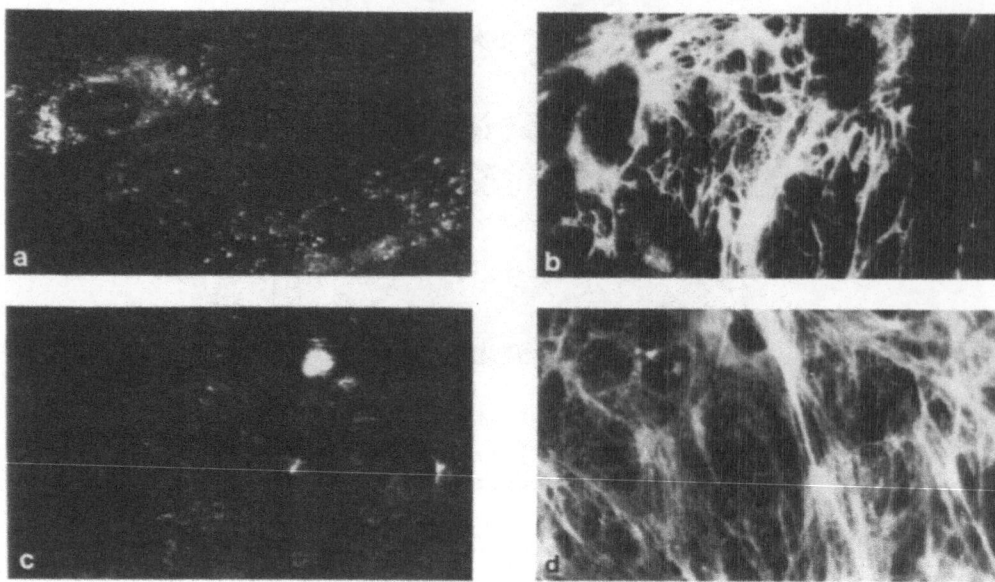

Fig. 2.   The progressive formation of fibronectin extracellular
          matrix in cultured endothelial cells.   Indirect immuno-
          fluorescence as in Fig. 1.   200 X.   For legends see
          text.

cells treated with protein synthesis inhibitors lack the fluorescent
intracellular granules (34,55).   While inside the cell fibronectin
quickly dimerizes.

     Fibronectin is present on endothelial cell surface, but it
is not yet clear if this fibronectin is associated with the extracellular matrix or bound to the plasma membrane.   It seems, however,
that this cell surface fibronectin does not possess the characteristics of an integral membrane protein (46).   The distribution
of fibronectin on cell surface is fibrillar (Fig. 1), and it seems
that it has a transmembrane linkage to the cytoskeleton.   Intracellular actin filaments align with the extracellular fibronectin
fibrils, and it has been shown that there is a direct correlation
between fibronectin concentration and the degree of intracellular
microfilament organization (32).   It, therefore, seems that there
is a strong association between fibronectin and actin, and that
this association can regulate the organization of the cell cytoskeleton (23).

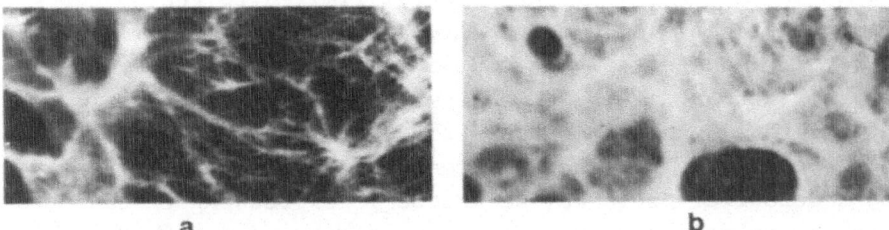

a   b

Fig. 3.   Culture-produced subendothelium from control endothelial
          cells (a), and from cells treated with dexamethasone (b).
          Dexamethasone induces the formation of a ticker extra-
          cellular matrix.   Immunofluorescence for fibronectin
          as in Fig. 1.   200 X.

Extracellular fibronectin accumulates in the cell layer of
cultured endothelial cells.   In the extracellular matrix, fibronectin
seems to act as a "molecular glue" that organizes and anchors the
components of connective tissue and basement membrane (31).   Fibronectin is an ubiquitous companion of collagen in reticulin fibers.
Endothelial cells show a marked polarity in synthesis of basement
membrane components, most of them being excreted towards the supporting layer (54).   Specific binding of fibronectin to collagen
has been demonstrated (9).   Factor XIIIa can also covalently
cross-link fibronectin to collagen (36).   The collagen-binding and
collagen-cross-linking site are separate.   Fibronectin can also
bind proteoglycans (heparin, heparan sulphate) (1).   Fibronectin
plays a major role in the extracellular matrix in the adhesion of
endothelial cells to Petri dishes and to collagen-coated surfaces
(38).   It also mediates cell-to-cell interactions.   It seems that
fibronectin forms the actual contact site between the cell and the
substratum, and functions as a substrate itself for cell attachment
and as a scaffold for cell migration and movement (15).

It is possible to follow the progressive in vitro formation
of the fibronectin extracellular matrix, which in cultured endothelial cells shows some peculiarities (42).   Few hours after the
beginning of the culture, fibronectin shows a granular intracellular
pattern localized in perinuclear areas (Fig 2a).   During the
following days, a progressive increase of fibronectin fibers is
seen.   At the subconfluent stage, fibronectin is present as a
meshwork of overlapping extracellular fibers (Fig. 2b).   When the
cells reach the confluence as a monolayer, fibronectin is only

detected as a thin extracellular material limited to the spaces
between adjacent cells. No intracellular fibronectin is visible.
No fibronectin is detectable on top of the monolayer (Fig. 2c)
(25,42). A treatment of this monolayer with hypotonic solutions
associated with a detergent exposes the underlying matrix, which
appears to be extremely rich in fibronectin (Fig. 2d). In this
respect, endothelial cells may be unique in expressing fibronectin
at confluence only in the extracellular matrix underneath the
cellular monolayer (54). On the contrary, for instance, the
cellular fibronectin of fibroblasts is initially located underneath
and in between cells, while as cells become confluent, it forms an
extracellular matrix that completely surrounds the cells.

     Cell surface fibronectin appears to play a role in cell-cell
interaction, as well as in adherence of cultured cells to a substratum (22,38,57). In vivo, it may modulate microvascular integrity, vascular permeability, and wound repair (56). Data from
studies on the distribution of fibronectin indicate that it plays
an important role in the orientation and positioning of cells in
tissues, and in the regulation of cell migration in vivo. These
properties are essential in the endothelial healing, and the
formation of a fibronectin extracellular matrix seems to be an
important permissive factor in the re-endothelization process,
providing a provisional substratum for cell migration and proliferation (17,28,53). In particular, it has recently been demonstrated
that blood vessels synthesize fibronectin in vivo in association
with endothelial proliferation (8). The same authors demonstrated
that circulating fibronectin is not recruited into the matrix of
these proliferating cells, suggesting that there might be a direct
relationship between endothelial proliferation and the increased
production of fibronectin by these cells. The extracellular matrix
can in vivo also mediate the induction of cell proliferation and
differentiation (15,16).

     It has recently been demonstrated that it is possible to af-
fect the synthesis of fibronectin by human endothelial cells in
culture (41,42). A reduction or an increase of fibronectin
synthesis has its influence on endothelial cell morphology, growth
rate, and extracellular matrix (44). Particularly, substances
that determine an increase of the deposition of fibronectin in
the extracellular matrix from endothelial cells are responsible
for a more flattened morphology with an increase in surface area

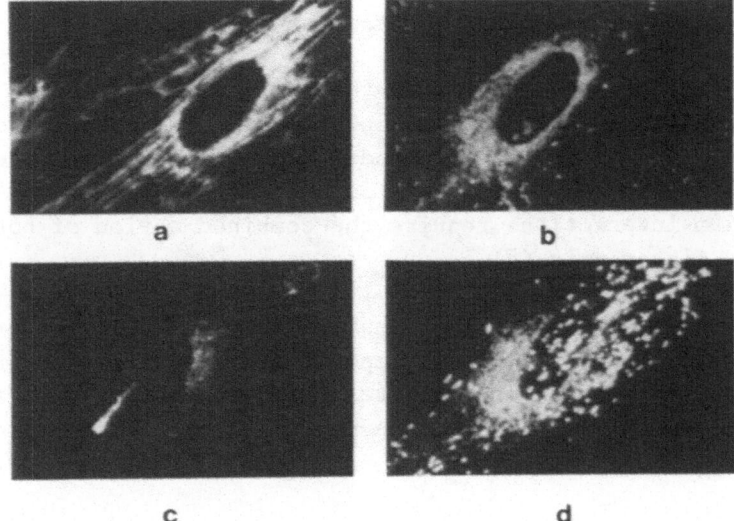

Fig. 4. An increase in fibronectin fibers is related
         to a decrease in VIIIR:Ag intracellular
         granules. Indirect combined immunofluorescence
         on endothelial cells grown in the presence
         of dexamethasone (a,b) or ticlopidine (c,d).
         Same cell preparations. 250 X.
         a,c: immunofluorescence for fibronectin (FITC).
         b,d: immunofluorescence for VIIIR:Ag (TRITC).

and for a more rapid repair after <u>in vitro</u> denudation, while the
opposite effect is seen with substances that dramatically decrease
the availability of fibronectin in culture dishes (Fig. 3).

In the same studies the existence of a close relationship,
i.e., an interdependence between fibronectin and VIIIR:Ag in
endothelial cell metabolism has been suggested (41).

Fibronectin and VIIIR:Ag share many similarities: both proteins
have the same subunit type of structure, both are present in a
secretory pool in platelets and in plasma, both are synthesized
by endothelial cells, and both have affinity for collagen. Moreover,
a close relationship between fibronectin and extracellular VIIIR:Ag
fibers in endothelial cells cultures has been demonstrated (43).
To date, it has been envisaged that substances that increase fibronectin content and synthesis in human endothelial cells seem to
decrease VIIIR:Ag content and synthesis in the same cells, while

the opposite effect is seen with substances that reduce fibronectin content and synthesis by human endothelium (Fig. 4).

It has recently been confirmed that extracellular matrix fibronectin has a role in platelet spreading in the overall sequence of events involved in platelet-subendothelium interactions and thrombus formation, and that the efficiency of these platelet-subendothelium interactions require the combined action of both VIIIR:Ag and fibronectin (5).

If all these suggested relationships between these two proteins will be confirmed in vivo, a new, important role in relation to hemostasis and thrombosis will be attributed to fibronectin, a protein that just a few years ago seemed only to be a molecule in search of a function.

REFERENCES

1.   S.K. Akiyama, K.M. Yamada, and M. Hayashi, The structure of fibronectin and its role in cellular adhesion, J. Supramol. Struct., 16:345 (1981).

2.   B.T. Atherton and R.O. Hynes, Differences between plasma and cellular fibronectin located with monoclonal antibodies, Cell, 25:133 (1981).

3.   H.B. Bensusan, T.L. Koh, K.G. Henry, B.A. Murray, and L.A. Culp, Evidence that fibronectin is the collagen receptor on platelet membrane, Proc. Natl. Acad. Sci. USA, 75:5864 (1978).

4.   C.R. Birdwell, D. Gospodarowicz, and G.L. Nicolson, Identification, localization, and role of fibronectin in cultured bovine endothelial cells, Proc. Natl. Acad. Sci. USA, 75:3273 (1978).

5.   F.M. Booyse, S. Fedre, and A.J. Quarfoot, Culture-produced subendothelium. II. Effect of plasma, FVIII:Wf and fibronectin on interactions of normal platelets with normal and von Willebrand porcine aortic subendothelium, Thromb. Res., 28:299 (1982).

6.   F.M. Booyse, B.J. Sedlak, and M.E. Rafelson, Culture of arterial endothelial cells: characterization and growth of bovine aortic cells, Thromb. Diath. Haemorr., 34:825 (1975).

7.   V. Buonassisi and J.C. Venter, Hormone and neutrotransmitter receptors in an established vascular endothelial cell line, Proc. Natl. Acad. Sci. USA, 73:1612 (1976).

8. R.A.F. Clark, J.H. Quinn, H.J. Winn, J.M. Lanigan, P. Delle-
    pella, and R.B. Colvin, Fibronectin is produced by blood
    vessels in response to injury, J. Exp. Med., 152:646 (1982).
9. E. Engvall, E. Ruoslahti, and E.J. Miller, Affinity of fibro-
    nectin to collagen of different genetic types and to
    fibrinogen, J. Exp. Med., 147:1584 (1978).
10. L.T. Furcht, D.F. Mosher, and G. Wedelschafer-Crabb, Effects
    of cell density and transformation on the formation of
    a fibronectin extracellular filamentous matrix on human
    fibroblasts, Cancer Res., 38:4618 (1978).
11. L.T. Furcht, D.F. Mosher, and G. Wendelschafer-Crabb, Immuno-
    cytochemical localization of fibronectin (LETS protein)
    on the surface of L6 myoblasts: light and electron
    microscopic studies, Cell, 13:263 (1978).
12. M.A. Gimbrone, Culture of vascular endothelial cells, in:
    "Progress in Haemostasis and Thrombosis," T.H. Spaet, ed.,
    Grune and Stratton, New York, Vol. 3 (1976).
13. J.C. Giddings, L.R. Brookes, F. Piovella, and A.L. Bloom,
    Immunohistological comparison of platelet factor 4 (PF4),
    fibronectin (Fn), and factor VIII-related antigen (VIIIR:Ag)
    in human platelet granules, Br. J. Haematol., 52:79 (1982).
14. M.H. Ginsberg, R.G. Painter, C. Birdwell, E.F. Plow, and J.
    Forsyth, The detection immunofluorescent localization,
    and thrombin-induced release of human platelet-associated
    fibronectin antigen, J. Supramol. Struct., 11:167 (1979).
15. D. Gospodarowicz, I. Vlodavsky, and N. Savion, The extracel-
    lular matrix and the control of proliferation of vascular
    endothelial and vascular smooth muscle cells, J. Supramol.
    Struct., 13:339 (1980).
16. D. Gospodarowicz, I. Vlodavsky, and N. Savion, The extracel-
    lular matrix, Endocrine Rev., 1:20 (1980).
17. F. Grinnell, R.E. Billingham, and L. Burgess, Distribution of
    fibronectin during wound healing in vivo, J. Invest.
    Dermatol., 76:181 (1981).
18. F. Grinnell and M.K. Feld, Distribution of fibronectin on
    peripheral blood cells in freshly clotted blood, Thromb.
    Res., 24:397 (1981).
19. F. Grinnell and D.G. Hays, Cell adhesion and spreading factor:
    similarity to cold insoluble globulin in human serum,
    Exp. Cell Res., 115:221 (1978).
20. K. Hedman, A. Vaheri, and J. Wartiovaara, External fibronectin
    of cultured human fibroblasts is predominantly a matrix
    protein, J. Cell Biol., 76:748 (1978).

21. B.J. Howard, E.J. Macarak, D. Gunson, and N. Kefalides,
    Characterization of the collagen synthesized by endothelial
    cells in culture, Proc. Natl. Acad. Sci. USA, 73:2361
    (1976).

22. R.O. Hynes, I.V. Ali, and A.J. Destree, A large glycoprotein
    lost from the surface of transformed cells, Ann. N.Y.
    Acad. Sci., 312:317 (1978).

23. R.O. Hynes and A. Destree, Relationship between fibronectin
    (LETS protein) and actin, Cell, 15:875 (1978).

24. E.A. Jaffe and D.F. Mosher, Synthesis of fibronectin by
    cultured human endothelial cells, Ann. N.Y. Acad. Sci.,
    312:122 (1979).

25. E.A. Jaffe and D.F. Mosher, Synthesis of fibronectin by
    cultured human endothelial cells, J. Exp. Med., 147:1779
    (1978).

26. E.A. Jaffe, R.L. Nachman, A.G. Becker, and C.R. Minick, Culture
    of human endothelial cells derived from umbilical veins.
    Identification by morphologic and immunologic criteria,
    J. Clin. Invest., 52:2745 (1973).

27. F. Jilek and H. Hörmann, Fibronectin (cold insoluble globulin).
    V. Mediation of fibrin-monomer binding to macrophages,
    Hoppe-Seyler's Z, Physiol. Chem., 359:1603 (1978).

28. M. Kurkinon, A. Vahery, P.J. Roberts, and S. Steinman,
    Sequential appearance of fibronectin and collagen in
    experimental granulation tissue, Lab. Invest., 43:47 (1980).

29. J. Lahav and R.O. Hynes, Involvement of fibronectin, von
    Willebrand factor, and fibrinogen in platelet interaction
    with solid substrata, J. Supramol. Struct., and Cell.
    Biochem., 17:299 (1981).

30. E.J. Macarak, E. Kirby, T. Kirk, and N.A. Kefalides, Synthesis
    of cold-insoluble globulin by cultured calf endothelial
    cells, Proc. Natl. Acad. Sci. USA, 75:2621 (1978).

31. J. McDonagh, Fibronectin. A molecular glue, Arch. Pathol.
    Lab. Med., 105:393 (1981).

32. C.J. Marshall, K.C. Humphryes, and R.E. Pollack, Microfilaments
    bundles, LETS protein, and growth control in somatic
    cell hybrids, J. Cell. Sci., 33:191 (1978).

33. M.W. Mosesson and D.L. Amrani, The structure and biological
    activities of fibronectin, Blood, 56:145 (1980).

34. D.F. Mosher, Fibronectin, Progr. Haemost. Thromb., 5:111
    (1980).

35. D.F. Mosher and L.T. Furcht, Fibronectin: review of its

structure and possible functions, J. Invest. Dermatol.,
77:175 (1981).

36. D.F. Mosher, P.E. Schod, and H.K. Kleinman, Cross-linking of
     fibronectin to collagen by blood coagulation factor XIIIa,
     J. Clin. Invest., 64:781 (1979).

37. M.R. Owens and C.D. Cimino, Synthesis of fibronectin by the
     isolated perfused liver, Blood, 59:1305 (1982).

38. E. Pearlstein, Plasma membrane glycoprotein which mediates
     adhesion of fibroblasts to collagen, Nature, 262:497 (1976).

39. E. Pearlstein and L.I. Gold, High molecular weight glycoprotein
     as a mediator of cellular adhesion, Ann. N.Y. Acad. Sci.,
     312:278 (1978).

40. E. Pearlstein, L. Gold, and A. Garcia Pardo, Fibronectin: a
     review of its structure and biological activity, Mol. Cell.
     Biochem., 29-103 (1980).

41. F. Piovella, J.C. Giddings, P. Almasio, M.M. Ricetti, and E.
     Thomas, Effect of ticlopidine and dexamethasone on
     fibronectin and factor VIII-related antigen synthesis by
     cultured endothelial cells, Thromb. Res., in press.

42. F. Piovella, J.C. Giddings, M.M. Ricetti, P. Almasio, and E.
     Ascari, The effect of dexamethasone on formation of intra-
     and extracellular fibronectin matrix by human endothelial
     cells in culture, Haematologica, 67:58 (1982).

43. F. Piovella, J.C. Giddings, and J.E. Thomas, Factor VIII-related
     antigen, fibronectin, and collagen type III and IV in
     cultured endothelial cells and in culture-derived sub-
     endothelium. An immunofluorescent study, Proceedings
     Symposium on Factor VIII-von Willebrand Factor, San Diego,
     Ca., 7th-9th October (1982).

44. F. Piovella, L. Piovella, P. Almasio, M.M. Ricetti and E.
     Ascari, The effect of ticlopidine on cultured endothelial
     cells: possible influence on extracellular matrix formation,
     Thromb. Haemostas., 46:404 (1981).

45. E.F. Plow, C.R. Birdwell, and M.H. Ginsberg, Identification
     and quantitation of platelet-associated fibronectin antigen,
     J. Clin. Invest., 63:540 (1979).

46. E. Ruoslahti, E. Engvall, and E. Hayman, Fibronectin: current
     concepts of its structure and functions, Coll. Res., 1:95
     (1981).

47. T.M. Saba, R.A. Blumenstock, P. Weber, and J.E. Kaplan,
     Physiologic role of cold-insoluble globulin in systemic
     host defense: implications of its characterization as the

opsonic α_2-surface binding glycoprotein, Ann. N.Y. Acad.,
Sci., 312:43 (1978).

48. T.M. Saba and E. Jaffe, Plasma fibronectin (opsonic glycoprotein): its synthesis by vascular endothelial cells
and role in cardiopulmonary integrity after trauma as
related to reticuloendothelial function, Am. J. Med.,
68:577 (1980).

49. S.A. Santoro and L.W. Cunningham, Fibronectin and the multiple
interaction model for platelet-collagen adhesion, Proc.
Natl. Acad. Sci. USA, 76:2644 (1979).

50. D.N. Slater and J.M. Sloan, The porcine endothelial cell in
tissue culture, Atherosclerosis, 21:259 (1975).

51. A. Vaheri, K. Alitalo, and K. Hedman, Fibronectin and the
pericellular matrix of normal and transformed adherent
cells, Ann. N.Y. Acad. Sci., 312:343 (1978).

52. A. Vaheri and D.F. Mosher, High molecular weight, cell surface-
associated glycoprotein (fibronectin) lost in malignant
transformation, Biochem. Biophys. Acta, 516:1 (1978).

53. I. Vlodavsky, A. Eldor, E. Hyam, R. Atzmon, and Z. Fuks,
Platelet interaction with the extracellular matrix produced
by cultured endothelial cells: a model to study the
thrombogenicity of isolated subendothelial basal lamina,
Thromb. Res., 28:179 (1982).

54. B. Waxler, B. Schumacher, and R. Eisenstein, Cell stroma
interactions in aortic endothelial cell cultures, Lab.
Invest., 41:128 (1979).

55. K.M. Yamada, Immunological characterization of a major trans-
formation-sensitive fibroblast cell surface glycoprotein.
Localization, redistribution, and role in cell shape,
J. Cell. Biol., 73:520 (1978).

56. K.M. Yamada, Fibronectin and other structural proteins, in:
"Cell Biology of Extracellular Matrix," E.D. Hay, ed.,
Plenum Publ. Corp., New York (1981).

57. K.M. Yamada and K. Olden, Fibronectins: adhesive glycoproteins
of cell surface and blood, Nature, 275:179 (1978).

58. M.B. Zucker, M.V. Mosesson, M.J. Bockman, and K.L. Kaplan,
Release of platelet fibronectin (cold-insoluble globulin)
from alpha granules induced by thrombin or collagen:
lack of requirement for plasma fibronectin in ADP-induced
platelet aggregation, Blood, 54:12 (1979).

PLATELETS AND THE PLATELET-DERIVED GROWTH FACTOR

IN THE PATHOGENESIS OF ATHEROSCLEROSIS

Russell Ross

University of Washington
Department of Pathology
Seattle, Washington 98195

INTRODUCTION

Atherosclerosis is the most common cause of heart attack and stroke in western civilization. Although this disease process has been recognized for well over 100 years, the nature of the cellular constituents that comprise the lesions of atherosclerosis, the etiology of this disease process, and its course and progress have remained either unknown or highly controversial. The last decade or so has been most productive in providing extensive new information concerning the cellular content of the lesions of atherosclerosis, and has provided several interesting hypotheses concerning the pathogenesis of atherosclerosis that have permitted us to ask new questions, design new experiments, and develop an understanding of this disease, so that it should be possible in the not-too-distant future to provide improved diagnosis in patients at risk, and to develop means for both intervention and prevention of this disease process, and thus of both myocardial and cerebral infarction.

LESIONS OF ATHEROSCLEROSIS

Although atherosclerosis was originally thought to be a degenerative process, there is now general agreement that the principal lesion of atherosclerosis, the fibrous plaque, represents an intimal smooth muscle proliferative response consisting of large numbers of smooth muscle cells together with variable numbers of macrophages.

.

The smooth muscle cells in the lesions are surrounded by all of the
elements of the connective tissue including collagen, elastic fiber
proteins, and proteoglycans, all of which are synthesized by the
newly proliferated smooth muscle cells within the lesions themselves.
The lesions consist not only of proliferated intimal smooth muscle
cells and newly formed connective tissue, but dependent upon the
location of the artery involved and the age of the lesion, the
lesions will have variable amounts of intracellular and extracellular lipid consisting principally of cholesterol and cholesterol
esters.

Therefore, if we are to understand how the lesions of atherosclerosis occur, this will require an understanding of how one can
generate within the intima of a given artery, smooth mucle proliferation, lipid accumulation, and new connective tissue formation, since
these are the three key elements of each lesion.

The earliest lesion of atherosclerosis to be found in man is
the so-called "fatty streak." This lesion consists largely of
lipid-laden macrophages, derived from blood monocytes, and causes
little to no clinical problems. The fatty streak has been found in
very young individuals, from infants to young adolescents, and in
these individuals it is distributed throughout the arterial tree. In
contrast, the more advanced lesions of atherosclerosis, the fibrous
plaque and the so-called "complicated lesion," are proliferative
lesions consisting of numerous smooth muscle cells together with
variable numbers of macrophages. They are generally found at
branching sites, bifurcations, and other areas that are thought to
be under various kinds of stress, including the shear stress from
the flow of blood as well as other stresses on the arteries themselves. Thus, there is a difference in the distribution of the
fatty streak and the more advanced lesions. The explanation for
this difference in distribution has been attributed to the fact that
where the fatty streaks and the more advanced lesions coincide, this
represents progression from fatty streaks to the more proliferative
lesions. Where they do not coincide, it has been suggested that
fatty streaks represent a form of lipid infiltration in the artery
wall, and that whatever the source of interaction with the artery
wall at the time of the development of the fatty streak, this interaction may have ceased so that the lesion did not progress to the
more advanced form.

The genesis of these lesions of atherosclerosis has been explained by a hypothesis termed "the response to injury of atherosclerosis" that was formulated by my colleagues, John Glomset, Laurence Harker, and me, and that has been modified with time with the advent of new information concerning this disease process.

THE RESPONSE TO INJURY HYPOTHESIS OF ATHEROSCLEROSIS

This hypothesis of atherogenesis suggests that the initiating factors that lead to the formation of the lesions of atherosclerosis are represented by various forms of "injury" to the endothelial lining of the artery. The "injury" may be quite different depending upon the agent. For example, the "injury" that may result from chronic hypercholesterolemia may be quite different from that which results from smoking, hypertension, diabetes, or many of the other risk factors that have been epidemiologically associated with an increased incidence of atherosclerosis. "Injury" to the endothelial cells can cover a wide spectrum of events. This spectrum may range from very subtle changes in the endothelial cells, altering perhaps their permeability characteristics with no morphologic manifestations, to at the other end of the spectrum endothelial cell-cell detachment, and endothelial cell-connective tissue detachment, leading to desquamation of the endothelial cells into the blood stream and denudation of the artery wall to expose the underlying subendothelial connective tissue. When the endothelial denudation is sufficient, this could lead to opportunities for interactions among a number of blood components and the exposed subendothelial connective tissue, including platelets, blood monocytes, and a number of plasma constituents including plasma lipoproteins. The interaction of these various constituents, according to the response to injury hypothesis, would then lead to migration of smooth muscle cells from the media of the artery into the intima, to proliferation of these newly arrived intimal smooth muscle cells, and to the formation of a fibromusculoelastic proliferative smooth muscle lesion. Such a lesion would be considered to be a pre-atherosclerotic response. With continued insult to the artery, the hypothesis suggests that the proliferative response continues and is augmented by stimulation of formation of new connective tissue proteins by the proliferated smooth muscle cells and, if the individual involved is hypercholesterolemic as well, to the accumulation both intracellularly and extracellularly of lipids, leading to the eventual formation of the classical advanced lesion of atherosclerosis, the fibrous plaque. The hypothesis also suggests that if the insult or the injury to the endothelium ceases, and if the lesion has not become too extensive or calcified, the lesions may be reversible and, therefore, may regress. On the other hand, if the injury is chronic over a long period of time, the hypothesis suggests that the lesions may continue to progress until the vascular lumen becomes sufficiently small due to lesion progression or to thrombosis, so that the blood supply to a given vessel may be compromised sufficiently to lead to clinical sequelae, including angina, heart attack, and stroke, depending upon the artery involved.

THE ROLE OF THE PLATELET

In examining the response to injury hypothesis, a number of studies have been pursued to determine the potential role of the platelet and the monocyte/macrophage in the process of atherogenesis. A number of laboratories have clearly demonstrated that when the endothelium is injured in the artery wall one of the first events to occur is the adherence and aggregation of platelets at such sites, and the release from the platelets of the contents of their granules followed within 24-72 hours by smooth muscle migration from the media into the intima and, further, to intimal proliferation of these smooth muscle cells, leading to the formation of an intimal smooth muscle proliferative lesion. When injury to the endothelium is induced mechanically, for example, with an intra-arterial catheter, and superimposed upon this injury is a hypercholesterolemic state as well, then the proliferative lesions take on the appearance of classical lesions of atherosclerosis.

Furthermore, several laboratories have demonstrated that if the platelet response that follows injury to the endothelium is ablated, for example, with the use of antiplatelet sera, with pharmacologic agents that prevent platelet adherence and release, or by using animals that are genetically unable to permit their platelets to adhere at such sites of injury, then in the absence of a platelet response, there is little to no intimal smooth muscle proliferative response. These studies provided interesting and exciting data to suggest that platelets provided a substance or substances that played a critical role in inducing the intimal smooth muscle accumulation that precedes frank atherosclerosis.

The role of the monocyte/macrophage in this process also has been examined critically because, as indicated earlier, the earliest lesion of atherosclerosis, the fatty streak, consists largely of macrophages which later also become associated with an intimal smooth muscle proliferation. The question remains how do platelets and macrophages induce the proliferation of connective tissue cells such as smooth muscle and fibroblasts?

THE PLATELET-DERIVED GROWTH FACTOR

Concomitant with the discovery of the importance of platelets in the induction of lesions of experimental atherosclerosis, it was discovered that platelets contain an extremely potent mitogen, the platelet-derived growth factor. In fact, the platelet-derived growth factor represents one of the principal components of whole blood serum that is responsible for the induction of cell proliferation in culture when cells such as fibroblasts or smooth muscle are grown in culture medium containing whole blood serum. The platelet-derived growth factor has been isolated and highly purified in a number of laboratories, and is a most interesting protein. It is a highly cationic protein (pI approximately 9.8) with a molecular weight of approximately 30,000. It consists of two polypeptide chains that are disulfide bonded and which are necessary for biological activity. The molecule is so potent that approximately 1 ng/ml of culture media is sufficient to induce a logarhythmic growth of cultures of fibroblasts or smooth muscle cells.

The platelet-derived growth factor has been shown to induce DNA synthesis in cells such as fibroblasts or smooth muscle by binding to a specific high-affinity cell surface receptor. The K_d for binding of PDGF to its cell surface receptor is approximately 10^{-11}M. Cells such as fibroblasts contain from 30-50,000 receptors per cell, dependent upon the species and the tissue involved. Binding of PDGF to its specific high affinity cell surface receptor induces a host of cellular responses that occur within seconds to many hours after binding. For example, within seconds after binding of PDGF to its receptor, a series of phosphorylation reactions occur involving both the receptor for PDGF and several cytoplasmic proteins. These phosphorylations involve activation of a protein kinase that phosphorylates on tyrosine residues that are highly specific and are characteristic for the response to this growth factor. Similarly, within a very short period of time after binding of PDGF to its receptor, cells that have been treated with PDGF

rapidly take up phosphatidylinositol, and within 5-10 minutes show
a marked increase in the formation of diacylglycerol that is subsequently degraded by a diglyceride lipase leading to the formation
of free arachidonic acid. This formation of free arachidonic acid
that results from interaction with PDGF may represent the substrate
for formation of increased prostacyclin by smooth muscle cells that
is generated when smooth muscle cells are treated with PDGF.

In addition, PDGF stimulates increased protein and RNA synthesis, and increased binding of low-density lipoprotein to its high-
affinity cell surface receptor due to an increase in the number of
LDL receptors that result from exposure of the cells to PDGF.

Of particular interest was the observation by Grotendorst and
his colleagues that PDGF is a chemoattractant. Using a Boyden
chamber preparation, these investigators were able to demonstrate
that of all of the growth factors that have been purified thus far,
only PDGF was chemotactic for smooth muscle cells and fibroblasts.
This observation helps to potentially explain the in vivo data
showing that soon after endothelial denudation, platelet adherence
and release, smooth muscle cells were stimulated to migrate from
the media into the intima. This directional migration of the smooth
muscle cells may possibly be due to the chemotactic properties of
PDGF that may be released at the site where platelets adhere and
degranulate.

The observations noted above have been made in cell culture,
however, they suggest a host of responses of cells exposed to
platelet-derived growth factor ranging from initiation of DNA synthesis, which requires 18-24 hours at one end of the spectrum of
responses to very early responses such as increased phosphokinase
activation, phospholipid turnover, LDL binding and internalization,
cholesterol synthesis, protein and RNA synthesis, and finally
chemotaxis. All of these events potentially could play important
roles in the genesis of the lesions of atherosclerosis. What
remains to be proven, however, is whether PDGF is in fact the
mitogen that is causally related to the genesis of these events
in vivo.

Finally, Brown and his co-workers have demonstrated that
platelets play a critical role in the genesis of a new lesion of
atherosclerosis that occurs after coronary bypass surgery. It has

been shown that approximately 35% of all bypasses develop a new lesion of atherosclerosis at the perianastomotic site of the bypass (the patient's own saphenous vein) and the old coronary artery. A double-blind study was initiated by this group in which they studied the effects of aspirin and dipyridamole (a potent antiplatelet combination), aspirin alone (also a potent antiplatelet drug), and placebo. They found that both the aspirin and dipyridamole and the aspirin alone treated patients showed significant reductions in the formation of new lesions at these perianastomotic sites as compared with the placebo-treated patients. This is the first clinical evidence in man demonstrating the efficacy of an antiplatelet pharmacologic regimen in the prevention of formation of a new lesion of atherosclerosis.

THE MACROPHAGE

Macrophages may play a dual role in atherogenesis. They are well-known for their capacity for debridement in the removal of cell debris, lipids, etc. In 1976, we discovered that macrophages also make a growth factor, the macrophage-derived growth factor, that can also induce proliferation of connective tissue cells. This factor is a different protein from the one derived from the platelet. Its role, if any, in the process of atherosclerosis remains to be identified.

CONCLUSION

All of the observations above strongly suggest that platelets and one of their important products, the platelet-derived growth factor, play a critical role during or after endothelial injury in the pathogenesis of the lesions of atherosclerosis. The platelet-derived growth factor induces a host of biological responses in smooth muscle cells, all of which can be explained hypothetically to be important in various aspects of atherogenesis as they may be related to cell proliferation, lipid accumulation, and new connective tissue formation. Thus, all of these important events in atherosclerosis may be mediated by the platelet-derived growth factor through the interaction of platelets where endothelial cells are injured. Further research remains to be performed to test various aspects of this hypothesis. Nevertheless, sufficient data are already available to point to the development of new approaches in the diagnosis, treatment, and prevention of atherosclerosis.

REFERENCES

1. E. Bierman and R. Ross, Aging and atherosclerosis, in: "Athero-
      sclerosis Reviews," Vol. 2, R. Paoletti and A.M. Gotto,
      eds., Raven Press, New York, pp. 79-111 (1977).
2. D. Bowen-Pope and R. Ross, Platelet-derived growth factor. II.
      Specific binding on cultured cells, J. Biol. Chem., 257:
      5161-5171 (1982).
3. J. Chamley-Campbell, G.R. Campbell, and R. Ross, The smooth
      muscle cell in culture, Physiol. Rev., 59:1-61 (1979).
4. C. Gajdusek, P. DiCorleto, R. Ross, and S. Schwartz, An endo-
      thelial cell-derived growth factor, J. Cell Biol., 85:
      467-472 (1980).
5. K. Glenn, D. Bowen-Pope, and R. Ross, Platelet-derived growth
      factor. III. Identification of a PDGF receptor by af-
      finity labeling, J. Biol. Chem., 257:5172-5176 (1982).
6. A.J.R. Habenicht, J.A. Glomset, W.C. King, C. Nist, C.D.
      Mitchell, and R. Ross, Early changes in phosphatidyl-
      inositol and arachidonic acid metabolism in quiescent
      Swiss 3T3 cells stimulated to divide by platelet-derived
      growth factor, J. Biol. Chem., No. 23, 256:12329-12335
      (1981).
7. L. Harker, R. Ross, S. Slichter, and C. Scott, Homocystine-
      induced arteriosclerosis: the role of endothelial cell
      injury and platelet response in genesis, J. Clin. Invest.,
      58:731-741 (1976).
8. S.J. Leibovich and R. Ross, A macrophage-dependent factor that
      stimulates the proliferation of fibroblasts in vitro,
      Am. J. Pathol., 84:501-513 (1976).
9. E.W. Raines and R. Ross, Platelet-derived growth factor. I.
      High yield purification and evidence for multiple forms,
      J. Biol. Chem., 257:5154-5160 (1982).
10. R. Ross and J. Glomset, Atherosclerosis and the arterial smooth
      muscle cell, Science, 180:1332-1339 (1973).
11. R. Ross, J. Glomset, B. Kariya, and L. Harker, A platelet-
      dependent serum factor that stimulates the proliferation
      of arterial smooth muscle cell in vitro, Proc. Natl. Acad.
      Sci. U.S.A., 71:1207-1210 (1974).
12. R. Ross and J. Glomset, The pathogenesis of atherosclerosis,
      New Engl. J. Med., 295:369-377, 420-425 (1976).
13. R. Ross and A. Vogel, The platelet-derived growth factor, Cell,
      14:203-210 (1978).

14. R. Ross, The arterial wall and atherosclerosis, Annu. Rev. Med., 30:1-15 (1979).

15. R. Ross, George Lyman Duff Memorial lecture, Atherosclerosis - a problem of the biology of arterial wall cells and their interaction with blood components, Arteriosclerosis: A Journal of Vascular Biology and Disease, 1:293-311 (1981).

16. R. Ross, Smooth muscle cells and atherosclerosis, in: "Vascular Injury and Atherosclerosis," Marcel Dekker, New York, pp. 53-77 (1981).

17. R. Ross, Factors influencing atherogenesis, in: "The Heart," Fifth edition, J. Hurst, ed., McGraw-Hill, New York, pp. 935-950 (1981).

18. R.B. Rutherford and R. Ross, Platelet factors stimulate fibroblasts and smooth muscle cells quiescent in plasma serum to proliferate, J. Cell Biol., 69:196-203 (1976).

MONOCLONAL ANTIBODIES TO HUMAN VON WILLEBRAND FACTOR

STUDY OF ITS STRUCTURE-FUNCTION RELATIONSHIP

Dominique Meyer

Institut de Pathologie Cellulaire
Hôpital de Bicêtre
Paris, France

Immunochemistry and its application to quantitative and qualitative analysis of proteins of the hemostatic pathways has provided major advances in the measurement of coagulation proteins and their inhibitors, as well as related platelet receptor proteins, and in the basic understanding of molecular abnormalities in hereditary bleeding disorders. The recent development of cellular hybridization by Köhler and Milstein (1975,1976) for producing in virtually unlimited quantity hybridoma antibodies specific for a single antigenic determinant (epitope) on a molecule represents a revolutionary advance in immunochemistry. Such reagents permit immunologic assays of previously not achievable high specificity and reproducibility, identification of structural loci responsible for specific biological functions, purification of molecules otherwise refractory to the usual biochemical methods of isolation, and detailed probing of the surface of proteins.

In order to illustrate the principles of application of hybridoma antibodies to the molecular biology of the hemostasis proteins, the studies of human von Willebrand factor (vWF) will be addressed. This topic is of particular interest since monoclonal antibodies to this most complex protein have been employed to develop spatial maps of this protein, and illustrate its structure-function relationship.

vWF remains among the least well-understood proteins of the

hemostatic pathways though the topic of considerable study. This
glycoprotein is present in plasma at about 5-10 ug per ml as a set
of oligomers varying in molecular weight from 1 to 20 x 10^6 (Counts
et al., 1978; Fass et al., 1978; Meyer et al., 1980a; Ruggeri and
Zimmerman, 1980). It appears to be assembled from dimers of a
single 230,000 m.w. primary subunit (Count et al., 1978). vWF is
associated with factor VIII (VIII:C), and a number of lines of
evidence indicate that the two proteins are independent molecular
entities that are non-covalently linked.

Quantitation as well as characterization of the structure of
vWF has depended on immunochemical analyses. Antibodies specific
for vWF have been described in only a few humans with severe
homozygous von Willebrand's disease (vWD) who lack the protein
and have received vWF for therapy (Sarji et al., 1974; Mannucci
et al., 1976; Ruggeri et al., 1979). The latter antibodies are:
1) rare, 2) the immunochemical fine specificity has not been
defined nor is it necessarily identical from different individuals,
and 3) reactivity with factor VIII has not been unequivocally
excluded. Heterologous antisera have been produced by immunization
with relatively purified human vWF (Zimmerman et al., 1971). Such
antisera contain antibodies that neutralize ristocetin-cofactor
activity (VIIIR:RCo) and precipitate with specific antigenic
determinants (VIIIR:Ag). However, these antisera also neutralize
VIII:C to a variable degree, presumably due to the parallel induction of antibody responses to the VIII:C present in vWF. Although
of value, these heterologous antisera contain a heterogeneous mixture of antibodies to a diverse array of epitopes on this molecular
complex. Major problems in immunochemical studies of structure
thus result.

vWF has three recognized biological functions. First, it
appears to be essential for the adhesion of platelets to the
subendothelial components of vessel walls. Second, vWF is necessary for in vitro platelet aggregation induced by ristocetin
(VIIIR:RCo), an assumed parallel of the in vivo primary hemostatic
mechanism operative at the platelet surface. Third, vWF serves
as a carrier for factor VIII. vWF is synthesized by endothelial
cells (Jaffe et al., 1981). Clearly evident in the endothelial
cells (Bloom et al., 1973; Hoyer et al., 1973), it is also localized
in the subendothelium (Sakariassen et al., 1979; Rand et al.,
1980). It would appear that two forms of molecular traffic occur,

i.e., secretion into the plasma as well as deposition by the endothelial cell in the subattendant matrix of the vasculature. The
megakaryocyte is presently the only known alternative biosynthetic
source of vWF (Nachman et al., 1977). The presence of vWF in
platelets (Howard and Montgomery, 1974; Nachman and Jaffe, 1975)
presents a problem in distinguishing that carried from the megakaryocyte and packaged in the alpha granules (Nachman and Jaffe,
1975) from those molecules of vWF that may be absorbed from the
plasma. The alpha granule vWF is released by stimulation of
platelets with ADP, collagen or thrombin (Koutts et al., 1978),
a mechanism that could also play a role in facilitating the primary
hemostatic adhesion of platelets to the subendothelium.

Tschopp et al. (1974) first demonstrated the role of vWF in
platelet adhesion to the subendothelium using perfusion of rabbit
everted and denuded aorta with blood in the apparatus described by
Baumgartner (1973). Platelet adhesion is decreased using whole
blood from patients with severe vWD (Tschopp et al., 1974). A
similar defect is observed in normal blood preincubated with poly-
clonal anti-vWF antibodies (Baumgartner et al., 1980). Notably
the defect thus induced in platelet adhesion is corrected by adding
purified vWF to the blood (Weiss et al., 1978a). The defect in
platelet adhesion in the absence of vWF is most striking over a
defined range of high shear rate conditions (Weiss et al., 1978b;
Baumgartner et al., 1980) which are comparable to those found in
the microcirculation (Turitto and Baumgartner 1975 and 1979).

Major questions at the present include identification of the
intramolecular loci of vWF that mediate the various described
functions of this complex multimetric protein and the mechanisms
responsible for these functions. We have used a panel of hybridoma
antibodies to vWF (Meyer et al., 1980b) in an attempt to identify
the structural basis for platelet adhesion to the subendothelium
and ristocetin-induced platelet aggregation. These hybridomas were
produced in 1980 (Meyer et al., 1980b) and have remained stable
for three years. The specificity of these antibodies has been
substantiated by a variety of methods including binding to highly
purified vWF, absence of binding to fibrinogen, fibronectin and
plasma or cryoprecipitate from severe vWD, binding to different
size multimers of vWF in normal plasma, and binding to low molecular
weight multimers in type IIA vWD plasma. Monoclonality of the
original hybridomas has been substantiated by the typing of heavy

and light chains of antibodies. Since then, two other groups of
investigators have raised monoclonal antibodies to porcine (Katzmann
et al., 1981) or human (Sola et al., 1982) vWF.

The capacity of our monoclonal antibodies to influence the
adhesion of human platelets to the subendothelium or rabbit aorta
was analyzed in annular perfusion chambers (Meyer et al., 1981).
At a high shear rate (2,600 s^{-1}), similar to that of the microcirculation, two monoclonal antibodies added at 10 ug/ml to normal
citrated blood for 10 to 30 minutes at 37°C before perfusion
inhibited adhesion by 78% and 87% respectively. In contrast, no
inhibition was observed at low shear (200 s^{-1}), consistent with
observations testing blood from patients with vWD and indicating
that vWF plays little role in platelet adhesion at low shear flow.
Other hybridoma antibodies including ones specific for vWF as well
as for unrelated proteins were without effect at either shear rate.
Only one of these two hybridoma antibodies, capable of inhibiting
platelet adhesion, slightly attenuated VIIIR:RCo at 100 ug/ml.
A mixture of several monoclonal antibodies to vWF caused marked
inhibition of VIIIR:RCo, whereas, the same mixture had no effect
upon platelet adhesion. These observations are complemented by
the observation that one monoclonal antibody that inhibits VIIIR:RCo
had no effect upon platelet adhesion.

Our studies using monoclonal antibodies clearly indicate that
the events involved in the two main biological functions of vWF,
namely mediation of platelet adhesion to the subendothelium and
of ristocetin-induced platelet aggregation, are distinct and appear
not to involve the same loci on the surface of the molecule (Meyer
et al., 1981).

When our data are considered within the context of a hypothetical steric map of vWF surface based on steric hindrance
analysis, all monoclonal antibodies are each demonstrated to possess
independent epitope specificity. Thus, these monoclonal antibodies
permit the first surface, two-dimensional, map of vWF constructed
from competitive displacement data.

Considering that only two of the antibodies mildly inhibit
VIIIR:RCo at a high concentration (100 ug/ml), a hypothetical
placement of this site is proposed between these epitopes. In
addition, since the mixture of the monoclonal antibodies to vWF
displays more inhibitory effect, we suggest that the topographical

loci that are involved in ristocetin-mediated platelet-platelet interaction are cooperative. These loci have thus been tentatively localized in the center of vWF. Although there is no reason to assume that such a locus would exist in vivo, ristoceting being a non-physiological agent, the possibility that comparable interactions may participate in certain physiological functions cannot be excluded. The epitopes related to platelet adhesion to the subendothelium appear to be within moderate proximity of each other in that mild but reproducibly significant competitive displacement is observed between both IgG. Placement of future hybridoma antibodies within this spatial scheme and integration of the capacity of these antibodies to influence biological activities should progressively expand knowledge of the surface topography of vWF.

REFERENCES

Baumgartner, H.R., 1973, The role of blood flow in platelet adhesion, fibrin deposition and formation of mural thrombi, Microvas. Res., 5:167-179.

Baumgartner, H.R., Tschoop, T.B., and Meyer, D., 1980, Shear rate-dependent inhibition of platelet adehsion/aggregation on collagenous surfaces by antibodies to human factor VIII/von Willebrand Factor, Br. J. Haematol., 44:127-139.

Bloom, A.L., Giddings, J.C., and Wilks, C.J., 1973, Factor VIII on the vascular intima: possible importance in haemostasis and thrombosis, Nature New Biol., 241:217-219.

Counts, R.B., Paskell, S.L., and Elgee, S.K., 1978, Disulfide bonds and the quaternary structure of factor VIII/von Willebrand Factor, J. Clin. Invest., 62:702-709.

Fass, D.N., Knutson, G.J., and Bowie, E.J.W., 1978, Porcine Willebrand Factor: a population of multimers, J. Lab. Clin. Med., 91:307-320.

Howard, M.A., and Montgomery, D.C., 1974, Factor VIII related antigen in platelets, Thromb. Res., 4:617-623.

Hoyer, L.W., de los Santos, R.P., and Hoyer, J.R., 1973, Anti-hemophilic factor in endothelial cells by immunofluorescent microscopy, J. Clin. Invest., 52:2737-2744.

Jaffe, E.A., Hoyer, L.W., and Nachman, R.L., 1973, Synthesis of anti-hemophilic factor antigen by cultured human endothelial cells, J. Clin. Invest., 52:2754-2764.

Jaffe, E.A., Hoyer, L.W., and Nachman, R.L., 1974, Synthesis of von Willebrand factor by cultured human endothelial cells, Proc. Natl. Acad. Sci. USA, 71:1906-1909.

Jones, T.R., Kao, K.J., Pizzo, S.V., and Bigner, D.D., 1981, Endo-
    thelial cell surface expression and binding of factor VIII/
    von Willebrand Factor, Am. J. Pathol., 103:304-308.

Katzmann, J.A., Mujwid, D.K., Miller, R.S., and Fass, D.N., 1981,
    Monoclonal antibodies to von Willebrand Factor: reactivity
    with porcine and human antigen, Blood, 58:530-536.

Köhler, G., and Milstein, C., 1975, Continuous cultures of fused
    cells secreting antibody of predefined specificity, Nature,
    256:495-497.

Köhler, G., and Milstein, C., 1976, Derivation of specific antibody-
    producing tissue culture and tumor lines by cell fusions,
    Eur. J. Immunol., 6:511-519.

Koutts, J., Walsh, P.N., Plow, E.F., Fenton, J.W., Bouma, B.N., and
    Zimmerman, T.S., 1978, Active release of human platelet
    factor VIII related antigen by adenosine diphosphate, col-
    lagen, and thrombin, J. Clin. Invest., 62:1255-1263.

Mannucci, P.M., Meyer, D., Ruggeri, Z.M., Koutts, J., Ciavarella,
    N., and Lavergne, J.M., 1976, Precipitating antibodies in
    von Willebrand's disease, Nature, 262:141-142.

Meyer, D., Obert, B., Pietu, G., Lavergne, J.M., and Zimmerman,
    T.S., 1980a, Multimeric structure of factor VIII/von Wil-
    lebrand Factor in von Willebrand's disease, J. Lab. Clin.
    Med., 95:590-602.

Meyer, D., Obert, B., Zimmerman, T.S., and Edgington, T.S., 1980b,
    Monoclonal antibodies specific for factor VIII from cellular
    hybrids, Circulation, 62:395, (Abstract).

Meyer, D., Baumgartner, H.R., and Edgington, T.S., 1981, Effect
    of hybridoma antibodies to human factor VIII/von Willebrand
    Factor on the adhesion of platelets to the subendothelium,
    Blood, 58:237, (Abstract).

Nachman, R.L., and Jaffe, E.A., 1975, Subcellular platelet factor
    VIII antigen and von Willebrand Factor, J. Exp. Med., 141:
    1101-1113.

Nachman, R.L., Levine, R., and Jaffe, E.A., 1977, Synthesis of
    factor VIII antigen by cultured guinea pig megakaryocytes,
    J. Clin. Invest., 60:914-921.

Rand, J.H., Sussman, L.A., Gordon, S.V., Chu, S.V., and Solomon,
    V., 1980, Localization of factor VIII related antigen in
    human vascular subendothelium, Blood, 55:752-756.

Ruggeri, Z.M., Ciavarella, N., Mannucci, P.M., Molinari, A.,
    Dammacco, J.M., Lavergne, J.M., and Meyer, D., 1979,
    Familial incidence of precipitating antibodies in von

Willebrand's disease. A study of four cases, J. Lab. Clin. Med., 94:60-75.

Ruggeri, Z.M., and Zimmerman, T.S., 1980, Variant von Willebrand's disease: characterization of two subtypes by analysis of multimeric composition of factor VIII/von Willebrand factor in plasma and platelets, J. Clin. Invest., 65:1318-1325.

Sakariassen, K.S., Bolhuis, P.R., and Sixma, J.J., 1979, Human blood platelet adhesion to artery subendothelium is mediated by factor VIII-von Willebrand factor bound to the subendothelium, Nature, 279:636-638.

Sarji, K.E., Stratton, R.D., Wagner, R.H., and Brinkhous, K.H., 1974, Nature of von Willebrand factor. A new assay and a specific inhibitor, Proc. Natl. Acad. Sci. USA, 71:2937-2941.

Sola, B., Avner, P., Sultan, Y., Jeanneau, C., and Maisonneuve, P., 1982, Monoclonal antibodies against human factor VIII molecule neutralize antihemophilic factor and ristocetin cofactor activities, Proc. Natl. Acad. Sci. USA, 79:183-187.

Tschopp, T.B., Weiss, H.J., and Baumgartner, H.R., 1974, Decreased adhesion of platelets to subendothelium in von Willebrand's disease, J. Lab. Clin. Med., 83:296-300.

Tuddenham, E.G.D., Lazarchick, J., and Hoyer, L.W., 1981, Synthesis and release of factor VIII by cultured human endothelial cells, Br. J. Haematol., 47:617-626.

Turitto, V.T., and Baumgartner, H.R., 1975, Platelet deposition on subendothelium exposed to flowing blood: mathematical analysis of physical parameters, Trans. Am. Soc. Artif. Inter. Organs, 21:593-600.

Turitto, V.T., and Baumgartner, H.R., 1979, Platelet interaction with subendothelium in flowing rabbit blood: effect of blood shear rate, Microvac. Res., 17:38-54.

Wall, R.T., Counts, R.B., Harker, L.A., Strikel, G.E., 1980, Binding and release of factor VIII/von Willebrand's factor by human endothelial cells, Br. J. Haematol., 46:287-298.

Weiss, H.J., Baumgartner, H.R., Tschoop, V.T., and Cohen, D., 1978a, Correction by factor VIII of the impaired platelet adhesion to subendothelium in the von Willebrand's disease, Blood, 51:267-279.

Weiss, H.J., Turitto, V.T., and Baumgartner, H.R., 1978b, Effect of shear rate on platelet in interaction with subendothelium in citrated and native blood. I. Shear depedent decrease of adhesion in von Willebrand's disease and the Bernard-Soulier syndrome, J. Lab. Clin. Med., 92:750-764.

Zimmerman, T.S., Ratnoff, O.D., and Powell, A.E., 1971, Immunologic
    differentiation of classic hemophilia (factor VIII deficiency)
    and von Willebrand's disease, J. Clin. Invest., 50:244-254.

THE MEMBRANE GLYCOPROTEINS:

ADHESION AND AGGREGATION OF PLATELETS

Edoardo Ascari, Carlo L. Balduini,
Fabiola Sinigaglia, and Cesare Balduini

Department of Internal Medicine and
Department of Biochemistry
University of Pavia
Italy

INTRODUCTION

Though the membrane of platelets, together with that of
erythrocytes, is one of the most studied from a biochemical and
functional point of view, that which we know about it is only a
minimal part of what remains to be learned. This situation is
largely due to the fact that platelets are very difficult cells to
study, since small differences in methodological approaches can
lead to quite opposite results. Surface glycoproteins, because of
their role in mediating platelet-platelet and platelet-extracellular
environment interactions, have recently received particular attention. In this paper, we attempt to summarize our knowledge about
their structure and function.

STRUCTURE OF MEMBRANE GLYCOPROTEINS

After hypotonic lysis of platelets, the membranes obtained by
density gradient centrifugation contain by weight 32-40% aminoacids, 7% carbohydrates, and 48-56% lipids represented by phospholipids (39-44%) and cholesterol (9-11%). By the use of electrophoretic techniques, we can identify about 30 glycoproteins in the
platelet membrane, but no more than four of them have been partially
characterized from a structural and functional point of view.

They are, according to the nomenclature of Phillips and Agin (1),
I_b, II_b, III, and V, and, according to Clemetson et al. (2), I_b,
II_b, III_a, and IV.

GLYCOPROTEIN I_b - GLYCOCALICIN

Glycoprotein I_b (GP I_b), the most prominent band after PAS
staining of SDS-polyacrylamide gel electrophoresis (SDS-PAGE), has
an apparent molecular weight of 170,000 (under non-reducing conditions); it consists of two subunits (Ib_α and Ib_β) linked by
disulfide bonds, with molecular weights of 143,000 and 22,000
respectively (1); both subunits are glycosylated and exposed on
the outer surface of membrane (1). GP I_b is not associated with
the cytoskeleton, either in the resting or in the thrombin activated
platelet (3), and it can be obtained from membrane by salt extraction (4).

When washed platelets are left for 24 hours in a buffer without
EDTA, GP I_b disappears from membrane, while a new glycoprotein
appears (5); the same glycoprotein is released from membrane when
platelets are lysed in the absence of EDTA or protease inhibitor (6).
This glycoprotein was named glycocalicin by Okumura and Jamieson
who first described it (7) and glycoprotein Is, because of its
water solubility, by Solum et al. (8). It is a single chain glycoprotein with a molecular weight of 148,000 (SDS-PAGE), containing
an intrachain disulfide bond and accounting for about 10% of total
platelet sialic acid (7). Carbohydrates represent 60% by weight
of its composition: galactose, N-acetylgamactosamine, N-acetyl-
glucosamine, and sialic acid are in a molar ratio of 2:1:1:2,
small amounts of glucose, mannose, and fucose being detectable (8).
Because the appearance of glycocalicin is inhibited by EDTA and
protease inhibitors, Solum et al. (6) suggested that it is derived
from GP I_b by the action of a calcium-dependent protease present in
platelets (9); recently Clemetson et al. (10) have confirmed this
hypothesis by the use of the tryptic fingerprint technique: tryptic
fragments of glycocalicin, though fewer, are in fact quite similar
to those obtained from GP I_b.

GLYCOPROTEINS II_b AND III

Glycoprotein II_b has an apparent molecular weight of 142,000
(SDS-PAGE) and consists of two subunits linked by disulfide bonds,

GP IIb$_\alpha$ and GP IIb$_\beta$, having molecular weights of 132,000 and 23,000 respectively. GP III is a single chain glycoprotein containing at least two intrachain disulfide bonds; its molecular weight is 114,000 when reduced and 99,000 in non-reducing conditions (SDS-PAGE) (11). Both glycoproteins are exposed on the outer membrane surface being labelled by lactoperoxidase-catalyzed iodination of intact platelets (12) and are embedded in the lipidic bilayer, being solubilized only after membrane disruption by detergents (2): for these reasons they are classified as integral glycoproteins.

While in resting platelets, these two membranes components do not seem to be bound to cytoskeleton. After thrombin-induced cell aggregation, they are extracted by Triton treatment as a complex with actin, myosin, and acting-binding protein (13). Because GP II$_b$ and GP III copurify with different methods of extraction, it has been recently hypothesized that GP II$_b$ and GP III could represent the subunits of a single glycoprotein (12); more recent evidence demonstrates that this is not true, because they have quite different tryptic peptide maps (14), aminoacid composition, and glycosylation (15). Polley et al. (16), using immunoelectron microscopy, have demonstrated that the two glycoproteins are distinct entities in resting platelets, but form clusters after thrombin stimulation. The ability of GP II$_b$ and III to form complexes is confirmed by Pidard et al. (17) and Gostad et al. (18) who demonstrated that this process is Ca^{++} or Mg^{++} dependent.

GLYCOPROTEIN V

Glycoprotein V has been recently isolated by Berndt and Phillips (19). It has a molecular weight of 82,000 (SDS-PAGE), both in reducing and non-reducing conditions, and contains 48% carbohydrate by weight, consisting of neutral hexose, hexosamine, and sialic acid in a molar ratio of about 8:2:1.

Because it can be extracted from the plasma membrane by varying the ionic strength, glycoprotein V can be classified as a peripheral protein. It is the sole membrane protein that is hydrolyzed by thrombin in intact platelets, releasing into the supernatant a fragment with a molecular weight of 69,500. It is interesting to note that normal plasma contains material anti-genically related to GP V; so it is not clear whether glycoprotein V is a plasma protein bound to platelets or a membrane molecule in

equilibrium with plasma, though it is also possible that the same
antigenic activity corresponds to different molecules.

MEMBRANE GLYOPROTEIN MODIFICATIONS DURING PLATELET AGING

Though the role of some of these surface glycoproteins seems
to be clear, it must be remarked that old platelets show a reduced
surface sialic acid content with respect to the younger ones, as
demonstrated by Rand et al. (20). Probably, this modification is
evidence that, during platelet life-span, a remodeling of membrane
glycoproteins occurs which could result from an autolytic phenomenon
similar to that we have observed during in vitro aging of intact
red cells (21) or isolated hosts (22). In our laboratory, it was
recently proven (23) that isolated platelet plasma membranes, after
in vitro incubation, are also significantly modified in their
glycoprotein composition (Tab. 1). Similar evidence was obtained
when washed platelets were incubated in a nutrient-free buffer over
a period of 24-28 hours. From this evidence, it seems reasonable
to suppose that circulating platelets are an heterogeneous population with respect to their surface glycoprotein composition; it is,
therefore, possible that the functions are also correlated with
these component changes during platelet life-span.

ROLE OF MEMBRANE GLYCOPROTEIN IN PLATELET ADHESION

Platelets have the ability to adhere and spread on exposed
subendothelium when the vessel endothelial layer has been disrupted. This event, representing the first step in the hemostatic
process, is very complex, probably involving different membrane
structures, subendothelium components, and plasma molecules. Two
congenital bleeding disorders, Bernard-Soulier (BS) syndrome and
von Willebrand's disease, represent a unique opportunity for the
investigation of this process. Bernard-Soulier platelets lack
GP I_b, bind less von Willebrand factor (vWF) to their membranes,
and show a decreased, but not absent, ability to adhere
to subendothelium (24); moreover, antibodies to GP I_b (25) induce
the functional defect of BS platelets in normal ones. These observations indicate that GP I_b plays an important role in platelet
adhesion, though deficient platelets can still adhere to some
extent to endothelium.

In von Willebrand's disease platelets also adhere to subendothelium poorly, even though their membranes contain normal amounts

Table 1. Hexosamine, Sialic Acid, and Aminoacid Content of Platelet
Membranes and their Modifications after in vitro Incubation in Krebs Ringer Phosphate Saline Buffer for 7 Hours
at 37° C

	Before Incubation	Modifications Induced by Incubation
	nmoles/mg Protein	nmoles/mg Protein
N—Ac. Glucosamine	39.0 + 11.9	+ 1.6 + 0.6
N—Ac. Galactosamine	7.5 + 2.0	− 2.3 + 0.4
Sialic Acid	6.3 + 1.8	− 2.7 + 0.5
	moles/100 moles	Variations %
Aspartic Acid	12.6 + 2.4	− 6.9 + 4.7
Threonine	5.2 + 0.2	− 6.3 + 4.2
Serine	9.5 + 2.1	− 1.6 + 6.4
Glutamic Acid	12.2 + 1.4	− 9.4 + 4.5
Proline	5.5 + 2.7	+ 3.8 + 19.6
Glycine	6.8 + 1.9	+ 10.4 + 13.6
Alanine	7.5 + 0.2	− 5.9 + 5.9
Valine	6.7 + 0.9	+ 12.0 + 6.8
Isoleucine	4.1 + 0.4	+ 23.9 + 16.3
Leucine	10.7 + 1.5	− 1.6 + 16.9
Tyrosine	traces	traces
Phenylalanine	3.8 + 0.3	+ 8.6 + 1.9
Histidine	2.0 + 0.4	+ 24.7 + 37.5
Lysine	6.7 + 0.7	− 6.2 + 9.5
Arginine	5.1 + 1.4	+ 3.6 + 6.6

Results are mean + S.D. of three experiments

of GP I_b. This function is restored by the addition of vWF from
normal subjects, indicating that vWF is involved in the mechanism
of platelet adhesion. Moreover, antibodies against vWF inhibit
platelet adhesion to subendothelium in Baumgartner's perfusion
chamber. It is interesting to note that the inhibition is strong
when platelet-subendothelium contact is short because of high blood
flow rate, but is moderate when blood flow rate is low (26). From
the above results it seems reasonable to hypothesize that platelets

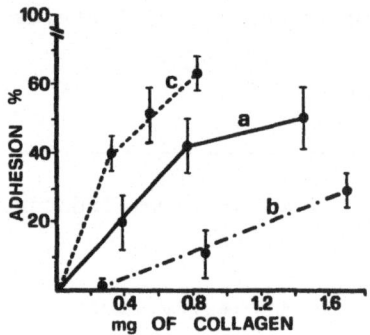

Fig. 1. Adhesion of platelets (1.5×10^8) to increasing amount (mg of collagen coupled to Sepharose) of different collagen preparations. a) pepsinized soluble collagen from bovine tendon. b) acid soluble collagen from calf skin. c) microfibrillar, pepsinized collagen from calf skin.

can adhere to the subendothelium by at least two mechanism, one of them involving vWF and GP I_b.

ADHESION INVOLVING GP I_b

In the presence of the antibiotic ristocetin normal platelets bind a large amount of vWF to their surface, while BS platelets, lacking GP I_b, do so to a lesser extent. Moreover, monoclonal antibodies to GP I partially inhibit ristocetin-induced vWF binding to membrane, indicating that GP I_b could represent one of the platelet receptors for vWF (25). Ristocetin is a non-physiologic agent, but recent reports indicate that ADP (27) and thrombin, at concentrations that are generated during hemostatic process, can induce the binding of exogenous (28) or platelet derived vWF (29) to platelet surface. On the basis of the above considerations, it seems reasonable that the GP I_b-vWF complex has the function of receptor for subendothelium. The identity of the subendothelium component to which platelet links through the GP I_b-vWF complex is less clear. Ruan et al. (25), studying the effect of anti GP I antibodies on the adhesion of platelets to normal and collagenase-digested subendothelium, concluded that collagen does not seem to be involved in this process; so, a non-collagenous substance seems to be the target for the GP I_b-vWF complex. The function of subendothelial non-collagenous compounds in this regard is largely unknown, but a candidate for this function is the microfibril; in fact platelet membranes can adhere to microfibrillar glycoproteins only when vWF is present (30) and monoclonal antibodies to GP I_b completely inhibit microfibril-platelet interaction (25). In conclusion, the platelet GP I_b-vWF-subendothelial microfibril axis could represent one of the modes of platelet adhesion.

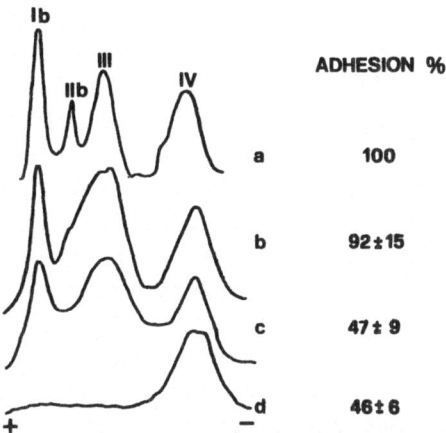

Fig. 2. Adhesion of surface modified human platelets to microfibril-
         lar collagen. a) normal platelets. b) platelets incubated
         at room temperature for 72 hours in plasma + CPD.
         c) platelets incubated at 37°C for 24 hours in Krebs Ringer
         phosphate saline, pH 7.4. d) platelets treated with
         trypsin (1 mg/10^9 cells) for 10', at 37°C. Left: SDS-PAGE
         of unreduced platelets performed by the method of Fairbanks
         et al. (32). PAS staining. Right: Adherence of washed
         platelets to columns of Sepharose 4B-microfibrillar col-
         lagen. The adhesion of surface modified platelets is
         expressed as % of the adhesion of normal platelets.

in agreement with those of Lahav and Meyer (33), who, using a different experimental model, also demonstrated that the reduction in
the ability of trypsin-treated platelets to adhere to collagen could
be attributed to the ADP released upon enzyme treatment and not to
the surface modifications; in fact, after addition of a high concentration of AMP, adhesion was restored to control values.

    Chiang and Kang (34) and Saito et al. (35) have recently
isolated two proteins that bind collagen from human and bovine
platelet membranes: they have similar molecular weights (65,000 in
human and 75,000 in bovine) and both compete with platelets in
binding to collagen. On the basis of these data, besides the GP I_b-
vWF-microfibril axis for platelet adhesion to subendothelium,
another one may be hypothesized, wherein this newly identified
membrane protein binds directly to collagen. Legrand et al. (36)
suggest that the receptor site on collagen could be a highly
repeating nonapeptide; in fact, it has been recently synthesized

and it is able to inhibit collagen-induced platelet aggregation. An interesting question is whether fibronectin, a glycoprotein present in platelet α-granules in plasma and in subendothelium and able to promote the adhesion and spreading of cells in tissue culture, also plays some role in the interaction between platelets and collagen. Many authors have studied this problem, but conflicting results have been obtained. Some came to the conclusion that fibronectin significantly enhances platelet adhesion to collagen also promoting the spreading of the cells (37), while others concluded that it plays no role (38). We also investigated this field using an affinity chromatography method, where washed platelet suspensions, some with and others without human plasma fibronectin, were passed through columns of Sepharose 4B-collagen to evaluate platelet retention, i.e., platelet adhesion. Our results (Fig. 3) indicate that the effect of fibronectin is different according to collagen preparations used: when collagen with high affinity for platelets is linked to Sepharose 4B, fibronectin stimulates platelet adhesion poorly; when collagen with low affinity for platelets is used, fibronectin significantly increases the extent of adhesion (31). Which of the two conditions physiologically occurs remains to be clarified.

Nevertheless, because fibronectin can bind to collagen (39) and also to a specific receptor recently identified in platelet membrane (40), a third axis for platelet adhesion to subendothelium may be hypothesized. It comprises the newly isolated membrane glycoprotein (m.w. 125,000), fibronectin, and collagen.

Preliminary results obtained in our laboratory seem to indicate that thrombin could also play a role in platelet adhesion to collagen. In fact, thrombin-treated platelets are retained to a larger extent by Sepharose 4B-collagen columns. This phenomenon is not dependent on platelet-release reaction and also occurs when proteolytic activity of thrombin is inhibited.

On the basis of the above considerations, we can conclude that if no definite mechanism of platelet adhesion to subendothelium is known, many can today be hypothesized. The reason for the existence of so many different types of adhesion could be that each of them becomes operative in the different anatomic and rheological situations that can be found in the vascular system.

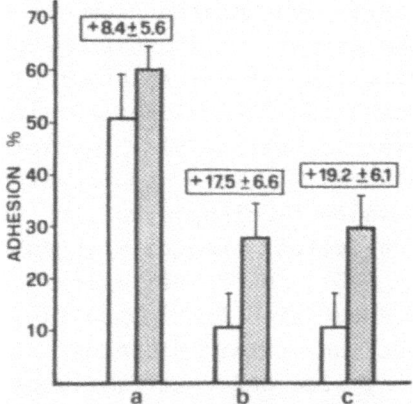

Fig. 3. Effect of fibronectin on the adhesion of platelets to
         Sepharose 4B-collagen columns. a) % adhesion of platelets
         ▨ not incubated or ☐ incubated with fibronectin to
         microfibrillar, pepsinized collagen from calf skin.
         b) % adhesion of platelets ☐ not incubated or ▨
         incubated with fibronectin to acid soluble collagen from
         calf skin. c) % adhesion of platelets to acid soluble
         collagen from calf skin ☐ not pre-treated or ▨ pre-
         treated with fibronectin. Results are mean + S.D. of
         three experiments.

ROLE OF MEMBRANE GLYCOPROTEINS IN PLATELET AGGREGATION

The molecular basis of platelet aggregation is less confusing
than that of platelet adhesion, as far as membrane glycoproteins
are concerned. The first in the explanation of this cellular event
derived from the study of platelets in Glanzmann's thrombasthenia.
In this inherited bleeding disorder, platelets adhere normally to
exposed subendothelium, undergo a release reaction after thrombin
stimulation, change shape in response to ADP indicating normal
surface receptors for physiological stimuli, but fail to aggregate
upon these stimuli. Nurden and Caen in 1974 (41) demonstrated an
important membrane modification in thrombasthenic platelets, i.e.,
the absence of one of the major glycoproteins. Subsequent studies,
using two dimensional electrophoresis, better characterized this
defect, consisting of the reduction of the amount of GP II$_b$ and
GP III (42). On the basis of these findings, it was hypothesized
that these glycoproteins could represent the aggregation site of
platelet membrane. This hypothesis was also supported by the

observation that an antibody isolated from the plasma of a poly-
transfused thrombasthenic patient inhibited ADP, collagen, and
thrombin-induced aggregation of normal human platelets (43). Since
extracellular fibrinogen is required for ADP-induced aggregation
(44), this molecule was thought to play the role of bridging protein
between stimulated platelets. This hypothesis has been sustained by
the observation that while fibrinogen has low affinity for un-
activated platelets, platelets treated with ADP show increased
binding of fibrinogen (45). Moreover, stimulated thrombasthenic
platelets fail to interact with fibrinogen (46), suggesting that
GP II$_b$ and GP III, lacking in this pathological condition, have
the function of binding site and that fibrinogen-GP II$_b$ and GP III
association is essential for cell aggregation. A recent work of
Gogstad (18) has confirmed this finding, indicating also that GP
II$_b$ and GP III have receptor function only in the form of divalent
cation mediated complex (see above), the single glycoproteins not
interacting with fibrinogen.

Thrombospondin also seems to be involved in aggregation,
since Jaffe et al. (47) have recently shown that this α-granule
glycoprotein, after secretion from platelet, is able to interact
with fibrinogen that has been linked to the cell surface.

In conclusion, we can today hypothesize that the following
molecular events lead to platelet aggregation. The exogenous
stimulus, reacting with platelet membrane, induces the formation
of the GP II$_b$-GP III complex, fibrinogen binds to the complex and
finally thrombospondin, available in plasma after release reaction,
reacts with fibrinogen on the surface of two adjacent cells
inducing aggregation.

ROLE OF MEMBRANE GLYCOPROTEINS IN THROMBIN-INDUCED ACTIVATION

The structure and function of platelet membrane is dramatically changed after thrombin stimulation. GP I displays receptor
activity for vWF (12), the complex between GP II$_b$ and GP III is
formed with receptor activity for fibrinogen (18), and cytoskeleton
binds to membrane through interaction with GP II$_b$-GP III (13).
We only know the first step of the mechanism by which thrombin
induces such modifications: it consists in the interaction with
the platelet membrane. Thrombin has two receptor sites on platelet
surface, represented by GP V and GP I$_b$. The former is hydrolyzed

Table 2. Adhesion of Human Washed Platelets to Columns of Sepharose 4B-Fibrillar Collagen

	Adhesion %
Platelets resuspended in buffer	25 ± 11
Platelets resuspended in fresh plasma	26 ± 3
Platelets preincubated with soluble collagen	3 ± 2

ADHESION NOT INVOLVING GP I_b

Despite a lack of GP I_b or the absence of vWF, platelets still adhere to some extent to subendothelium. It is, therefore, evident that some mechanism of adhesion to subendothelium different from that described above exists. Much evidence suggests that the sub-endothelium target for this type of adhesion is represented by collagen. One finding is that, in our experience (31), washed human platelets resuspended in medium not containing vWF can adhere to columns of Sepharose 4B-collagen; the extent of adhesion depends on the type of collagen organization (Fig. 1), and is unchanged when platelets are resuspended in fresh plasma before affinity chromato-graphy (Tab. 2).

Therefore, platelet surface can recognize and directly bind collagen; the membrane receptor seems to be the same for both soluble and fibrillar collagen, since preincubation of platelets with soluble collagen inhibits their subsequent adhesion to fibrillar collagen (Tab. 2).

Other evidence deriving from studies of adhesion to collagen of platelets modified in their membrane structure suggests that platelets can adhere to collagen independently of the presence of GP I_b in their membranes.

Figure 2 shows that thrombocytes incubated for different time in different medium, though having the same amount of GP I_b, differ in their ability to adhere to columns of Sepharose 4B-collagen. Moreover, trypsin-treated platelets, in which GP I_b is no more detectable, still partially adhere to collagen. These results are

by its action releasing a 69,500 molecular weight fragment, while
the latter is not (12). Which of the two receptors has the
functional role of stimulating platelets is not well known, but
much evidence suggests that it is represented by GP V. In fact,
antibodies against GP I$_b$ inhibit thrombin stimulated aggregation
poorly (48), BS platelets are activated by thrombin (49), while
several proteases that almost specifically hydrolyze GP V induce
platelet activation and, furthermore, thrombin hydrolysis of GP V
precedes platelet aggregation by few seconds (12). Obviously
the reported observations suggest, but do not prove, that GP V
is the functional receptor for thrombin and further investigations
are required to fully elucidate the role of membrane glycoproteins
in platelet activation.

ACKNOWLEDGEMENTS

This work was supported by grants from MPI and CNR - Italy

REFERENCES

1.   D.R. Phillips and P.P. Agin, Platelet plasma membrane glyco-
          proteins. Identification of a proteolytic substrate for
          thrombin, Biochem. Biophys. Res. Commun., 75: 940 (1977).
2.   K.J. Clemetson, S.L. Pfueller, E.F. Lüscher, and C.S.P.
          Jenkins, Isolation of the membrane glycoproteins of human
          blood platelets by lectin affinity chromatography,
          Biochem. Biophys. Acta, 464:493 (1977).
3.   A. Rotman, J. Heldman, and S. Linder, Association of membrane
          and cytoplasmatic proteins with the cytoskeleton in blood
          platelets, Biochemistry, 21:1713 (1982).
4.   T. Okumura, M. Hasitz, and G.A. Jamieson, Platelet glycocalicin.
          Interaction with thrombin and role as thrombin receptor
          of the platelet surface, J. Biol. Chem., 253:3435 (1978).
5.   E.F. Ali-Briggs, K.J. Clemetson, and C.S.P. Jenkins, Antibodies
          against platelet membrane glycoproteins, Br. J. Haematol.,
          48:305 (1981).
6.   N.O. Solum, I. Hagen, C. Filion-Myklebust, and T. Stabaek,
          Platelet glycocalicin: its membrane association and
          solubilization in aqueous media, Biochem. Biophys. Acta,
          597:235 (1980).
7.   T. Okumura and G.A. Jamieson, Platelet glycocalicin. Orienta-
          tion of glycoproteins of the human platelet surface,
          J. Biol. Chem., 251:5944 (1976).

8. N.O. Solum, I. Hagen, and M. Peterka, Human platelet glyco-
    proteins. Further evidence that "GP I band" from whole
    platelets contains three different polypeptides one of
    which may be involved in the interaction between platelets
    and factor VIII, Thromb. Res., 10:71 (1977).
9. D.R. Phillips and M. Jakabova, Ca$^+$-protease in human platelets,
    J. Biol. Chem., 252:5602 (1977).
10. K.J. Clemetson, H.Y. Naim, and E.F. Lüsher, Relationship
    between glycocalicin and glycoprotein Ib of human platelets,
    Proc. Natl. Acad. Sci. USA, 78:2712 (1981).
11. D.R. Phillips and P.P. Agin, Platelet plasma membrane glyco-
    proteins, J. Biol. Chem., 252:2121 (1977).
12. M.C. Berndt and D.R. Phillips, Platelet membrane proteins:
    composition and receptor functions, in: "Platelets and
    Biology and Pathology 2," J.I. Gordon, ed., Elsevier/
    North Holland Biomedical Press, Amsterdam (1981).
13. D.R. Phillips, L.K. Jennings, and H.H. Edwards, Identification
    of membrane protein mediating the interation of human
    platelets, J. Cell Biol., 86:77 (1980).
14. L.L.K. Lueg, T. Kinoshita, and R.L. Nachman, Isolation, Purifi-
    cation, and partial characterization of platelet membrane
    glycoproteins IIb and IIIa, J. Biol. Chem., 256:1994 (1981).
15. R.P. McEver, J.U. Baenziger, and P.W. Majerus, Isolation and
    structural characterization of the polypeptide subunits
    of membrane glycoprotein IIb-IIIa from human platelets,
    Blood, 59:80 (1982).
16. M.J. Polley, L.L.K. Lueg, F.Y. Clark, and R.L. Nachman, Thrombin
    induced platelet membrane glycoprotein IIb and IIIa complex
    formation, J. Exp. Med., 154:1058 (1981).
17. D. Pidard, J.P. Rosa, T.J. Kunicki, and A.T. Nurden, Further
    studies on the interaction between human platelet membrane
    glycoproteins IIb and IIIa in Triton X - 100, Blood, 60:
    894 (1982).
18. G.O. Gogstad, F. Brosstad, M.B. Krutnes, I. Hagen, and N.O.
    Solum, Fibrinogen-binding properties of human platelet
    glycoprotein IIb-IIIa complex: a study using crossed-
    radioimmunoelectrophoresis, Blood, 60:663 (1982).
19. M.C. Berndt and D.R. Phillips, Purification and preliminary
    physico-chemical characterization of human platelet membrane
    glycoprotein V, J. Biol. Chem., 256:59 (1981).
20. M.L. Rand, J.P. Greenberg, M.A. Packham, and J.F. Mustard,
    Density subpopulation of rabbit platelets: size, protein,
    and sialic acid content and specific radioactivity changes

following labeling with ^{35}S-sulfate _in vivo_, Blood, 57:741 (1981).

21. A. Brovelli, M. Suhail, G. Pallavicini, F. Sinigaglia, and C. Balduini, Self-digestion of human erythrocyte membranes. Role of adenosine triphosphate and glutathione, Biochem. J., 164:469 (1977).

22. A. Brovelli, G. Pallavicini, F. Sinigaglia, C.L. Balduini, and C. Balduini, Identification of a sialoglycopeptide released by self-digestion from human erythrocyte membranes, Biochem. J., 158:497 (1976).

23. F. Sinigaglia, C.L. Balduini, P. Salvini, M. Mazzucco, and C. Balduini, Changes in composition of platelet membranes after _in vitro_ incubation, It. J. Biochem., 31:428 (1982).

24. J.P. Caen, A.T. Nurden, C. Jeanneau, H. Michel, G. Tobleman, S. Levi-Toledano, Y. Sultan, F. Valensi, and J. Bernard, Bernard-Soulier syndrome: a new platelet glycoprotein abnormality. Its relationship with platelet adhesion to subendothelium and with the factor VIII von Willebrand protein, J. Lab. Clin. Invest., 87:586 (1976).

25. C. Ruan, G. Tobelem, A.J. McMichael, L. Drouet, Y. Legrand, L. Degos, N. Kieffer, H. Lee, and J.P. Caen, Monoclonal antibody to human platelet glycoprotein I, Br. J. Haematol., 49:511 (1981).

26. H.R. Baumgartner, T.B. Tschopp, and D. Meyer, Shear rate-dependent inhibition of platelet adhesion and aggregation on collagenous surfaces by antibodies to human factor VIII/von Willebrand, Br. J. Haematol., 44:127 (1980).

27. T. Fujimoto and J. Hawiger, Adenosine diphosphate induces binding of von Willebrand factor to human platelets, Nature, 297:154 (1982).

28. T. Fujimoto, S. Ohara, and J. Hawiger, Thrombin-induced exposure and prostacyclin inhibition of the receptor for factor VIII/von Willebrand factor on human platelets, J. Clin. Invest., 69:1212 (1982).

29. J.N. George, L.L. Thoi, L.M. McManus, and T.A. Reiman, Isolation of human platelet membranes microparticles from plasma and serum, Blood, 60:834 (1982).

30. Y. Legrand, F. Fauvel, N. Gutman, J.P. Muh, G. Tobelem, H. Souchon, H. Karniguian, and J.P. Caen, Microfibrils platelet interaction: requirement of von Willebrand factor, Thromb. Res., 19:737 (1980).

31. C.L. Balduini, F. Sinigaglia, P. Salvini, and C. Balduini, The role of plasma fibronectin in platelet adhesion to

collagen, Haemostasis, in press (1984).

32. J. Fairbanks, T.L. Steck, and D.H.F. Wallach, Electrophoretic analysis of the human erythrocyte membrane, Biochemistry, 10:2606 (1971).

33. J. Lavah and F.A. Meyer, On the role of the major platelet membrane glycoproteins in platelet adhesion to collagen, Thromb. Res., 22:457 (1981).

34. T.M. Chiang and A.H. Kang, Isolation and purification of collagen αl receptor from human platelet membrane, J. Biol. Chem., 257:7581 (1982).

35. Y. Saito, T. Imada, and Y. Ynada, Platelets adhering protein: isolation of a new non-fibronectin protein from bovine platelet membrane, Thromb. Res., 25:143 (1982).

36. Y.J. Legrand, A. Karmignan, P. Le Francier, F. Fauvel, and J.P. Caen, Evidence that a collagen-derived nonapeptide is a specific inhibitor of platelet collagen interaction, Biochem. Biophys. Res. Commun., 96:1579 (1980).

37. V.E. Koteliansky, V.L. Leytin, D.D. Sviridof, U.S. Repin, and V.N. Smirnov, Human plasma fibronectin promotes the adhesion and spreading of platelets on surfaces coated with fibrillar collagen, FEBS Letters, 123:59 (1981).

38. R.A. Sochynsky, B.J. Boughton, J. Burns, B.C. Sykes, and J. McGee, The effect of human fibronectin on platelet-collagen adhesion, Thromb. Res., 18:521 (1980).

39. D.F. Mosher, P.E. Schad, and H.K. Kleinman, Cross-linking of fibronectin to collagen by blood coagulation factor XIII_a, J. Clin. Invest., 64:781 (1979).

40. M.S. Hansen and I. Clemmensen, A fibronectin-binding glyco-protein from human platelet membranes, Biochem. J., 201: 629 (1982).

41. A.T. Nurden and J.P. Caen, An abnormal platelet glycoprotein pattern in three cases of Glanzmann's thrombasthenia, Br. J. Haematol., 28:253 (1974).

42. D.R. Phillips and P.P. Agin, Platelet membrane defects in Glanzmann's thrombasthenia, J. Clin. Invest., 60:535 (1977).

43. S. Levi-Toledano, G. Tobelem, G. Legrand, R. Bredoux, L. Degos, A. Nurden, and J.P. Caen, Acquired IgG antibody occurring in a thrombasthenic patient: its effect on human platelet function, Blood, 51:1065 (1978).

44. K.M. Brinkhous, M.S. Read, and R.G. Mason, Plasma thrombocyte agglutinating activity and fibrinogen synergism with adenosine diphosphate, Lab. Invest., 14:335 (1965).

45. J.F. Mustard, M.A. Packham, R.L. Kinlough-Rathbone, D.W. Perry,
     and E. Regoeezi, Fibrinogen and ADP-induced platelet ag-
     gregation, Blood, 52:453 (1978).
46. B.S. Coller, Interaction of normal, thrombasthenic, and Bernard-
     Soulier platelets with immobilized fibrinogen: defective
     platelet-fibrinogen interaction in thrombasthenia, Blood,
     55:168 (1980).
47. E.A. Jaffe, L.L.K. Leung, R.L. Nachman, R.I. Lewin, and D.F.
     Mosher, Thrombospondin is the endogenous lectin of human
     platelets, Nature, 295:246 (1982).
48. R.L. Nachman, E.A. Jaffe, and B.B. Weksler, Immunoinhibition
     of ristocetin-induced platelet aggregation, J. Clin.
     Invest., 59:143 (1977).
49. G.A. Jamieson and T. Okumura, Reduced thrombin binding and
     aggregation in Bernard-Soulier platelets, J. Clin. Invest.,
     61:861 (1978).

THE CHANGES IN ARACHIDONIC ACID METABOLISM IN ISCHEMIC

HEART DISEASE - THEIR PATHOPHYSIOLOGICAL CONSEQUENCES

G.G. Neri Serneri

Università di Firenze
Clinica Medica I
Florence, Italy

The interest in studying arachidonic acid (AA) metabolism in ischemic heart disease (IHD) comes from the awareness that prostacyclin (PGI_2) and thromboxane A_2 (TXA_2) are powerful modulators of vascular tone and platelet aggregability, and that they may be involved in the occurrence of various clinical manifestations of IHD. Platelets from IHD patients when stimulated produced a larger amount of TXB_2 in comparison with control platelets, and TXB_2 plasma levels in IHD patients were higher than the ones inhibited by ASA pretreatment (10 mg/kg i.v.). In contrast, plasma PGI_2-like activity and PGI_2-like activity produced by vessel wall after three-minute ischemia were significantly reduced in IHD patients (Neri Serneri et al., 1982). Moreover, in these patients an increase of TXB_2 plasma levels during ischemic attacks, symptomatic or asymptomatic, could be found even when TXB_2 formation by platelets was completely inhibited.

These findings suggest a disorder of AA metabolism in the platelets and vessel walls of IHD patients. By using high pressure liquid chromatography, we demonstrated that human arteries and veins produce TXA_2. Endothelial cells synthesize PGI_2, whereas, the media is able to produce significant amounts of TXA_2, in addition to PGI_2 (Neri Serneri et al., 1983). The PG's formed by vessel walls are released into the bloodstream after appropriate stimulation, and play an important role in modulating vascular response to adrenergic stimulation (Neri Serneri et al., 1981).

Adrenergic stimulation, obtained by cold application, induced a
sudden increase in plasma TXB_2 and PGI_2-like plasma activity in
arterial, venous, and coronary sinus blood. The increase in TXB_2
formation was not affected by ASA pretreatment (10 mg/kg), which
on the other hand completely inhibited the increase of plasma
PGI_2-like activity and TXA_2 formation by platelets. When adrenergic
stimulation was carried out in IHD patients, there was a significantly lower production of PGI_2-like activity than in the controls.
Increased formation of TXB_2 following adrenergic stimulation was
found in coronary sinus blood even after ASA administration.
Whereas, no significant differences could be found in relation to
the different clinical pictures of the disease (viz. effort angina,
spontaneous angina, and old myocardial infarction). Striking
differences in PGI_2 and TXA_2 production were found between patients
with inactive disease (viz. free from anginal attacks). In the
latter patients, PGI_2 production was even lower and TXB_2 formation
higher than in patients free from anginal attacks.

The imbalance in the production of TXA_2 and PGI_2-like activity
resulted in a significant increase in vascular resistance (measured
in the forearm) after adrenergic stimulation in comparison with
that found in controls and in patients free from anginal attacks.
Thus, the imbalance in vessel wall production of thromboxane and
prostacyclin results in impairment of the intravascular modulating
mechanism of the vascular response and, as a consequence, in
inappropriate contraction of the vascular wall to adrenergic stimulation. From the pathophysiological point of view, these abnormalities of the hemostatic system facilitate thrombus formation
and coronary vasospasm.

REFERENCES

Christ-Hazelhof, E., and Nugteren, D.H. 1981, Prostacyclin is not
        a circulating hormone, Prostaglandins, 22:739.
Haslam, R.Y., and McClenaghan, M.D., 1981, Measurement of circulating
        prostacyclin, Nature, 292:364.
Hensby, C.N., Fitzgerald, G.A., Friedman, L.A., Lewis, P.J., and
        Dollery, C.T., 1979, Measurement of 6-oxo-PGF_1 in human
        plasma using gas chromatography-mass spectrometry,
        Prostaglandings, 18:731.
Neri Serneri, G.G., Gensini, G.F., Abbate, R., Mugnaini, C.,
        Favilla, S., Brunelli, C., Chierchia, S., and Parodi, O.,

1981, Increased fibrinopeptide A formation and thromboxane A_2 production in patients with ischemic heart disease. Relationship with coronary pathoanatomy risk factors and clinical manifestations, Am. Heart J., 101:185.

Neri Serneri, G.G., Masotti, G., Gensini, G.F., Poggesi, L., Abbate, R., and Mannelli, M., 1981, Prostacyclin and thromboxane A_2 formation in response to adrenergic stimulation in humans: a mechanism for local control of vascular response to sympathetic activation? Cardiovas. Res., 15: 285.

Neri Serneri, G.G., Masotti, G., Poggesi, L., Galanti, G., Morettini, A., and Scarti, L., 1982, Reduced prostacyclin production in patients with different manifestations of ischemic heart disease, Am. J. Cardiol., 49:625.

Neri Serneri, G.G., Abbate, R., Gensini, G.F., Panetta, A., and Casolo, G.C., 1983, TXA_2 production by human arteries and veins, Prostaglandins, 19:754.

Steer, M.L., MacIntyre, D.E., Levine, L., and Salzman, E.W., 1980, Is prostacyclin a physiologically important circulating antiplatelet agent? Nature, 283:194.

HEMORHEOLOGY AND THROMBOSIS

J. Dormandy

St. James' & St. George's Hospitals
Sarsfeld Road
London, SW12, England

INTRODUCTION

Hemorheology is the study of the physical flow properties of
blood. Although one of the essential features of blood is that
it should flow, its rheological properties have only recently
received any attention. Thrombosis matters because it prevents
flow, but it may also be the result of changes in the hemorheological properties of blood. These are as important as its
respiratory, nutrative, hemostatic, and fibrinolytic properties.

The flow properties of blood are determined by a galaxy of
factors, represented in Figure 1. As will be seen later, these
interact in a complex manner. For instance, the plasma fibrinogen
not only largely determines the viscosity of the plasma alone,
but also affects whole blood viscosity by being the principal
factor causing red cell aggregability [1]. Plasma fibrinogen also
plays a role in platelet adhesiveness, and recently Wautier has
shown in an experimental model that plasma fibrinogen enhances
the adhesiveness of erythrocytes to endothelial cell cultures [2].
Red cell aggregability is thought to be the principal reason for
the increased viscosity of whole blood at low shear or flow rates.
Whilst at high shear rates, the deformability of the red cells is
of paramount importance [3]. The ability of the normal red cell
to deform is also vital to perfusion of the microcirculation,
where the caliber of the capillaries is usually much less than that

79

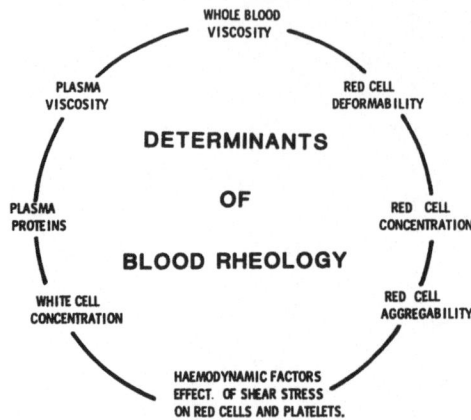

Fig. 1. Diagramatic represen-
         tation of the inter-
         action between the
         determinants of blood
         rheology.

of the undeformed red cells. Quantitatively the two most important
determinants of whole blood viscosity are simply the red cells and
the plasma fibrinogen concentrations. It is becoming increasingly
apparent that the mechanisms involved in these rheological phenomena
are basically the result of surface phenomena, either between red
cell and red cell or between red cell and the endothelium. These
surface phenomena are largely mediated by adsorbed proteins. The
whole picture is further complicated by the fact that not only does
the shear stress alter the viscosity of whole blood, but it also
seems to influence and be influenced by the red cell concentration
and the effect of shear forces on individual red cells; and platelets
may trigger biochemical reactions.

    For the most part, the assessment of the determinants of blood
rheology is now well standardized. Plasma and whole blood viscosity
at a range of shear rates are now routine measurements in hemorheological laboratories (4). Red cell aggregability is more difficult to quantify, but clinically the erythrocyte sedementation
rate is a reasonable approximation. The assessment of red cell
deformability is currently the subject of particular worldwide
interest, and Professor Di Perri will be dealing with this in the
next lecture (5).

    Conditions where hemorheological abnormalities have been

DISEASES ASSOCIATED WITH HAEMORHEOLOGICAL ABNORMALITIES

A. ABNORMAL PLASMA	B. HIGH RED CELL CONCENTRATION	C. ABNORMAL RED CELLS	
Macroglobulinaemia	Primary Polycythaemia	i Shape	{ Spherocytosis Macrocytosis
Cryoglobulinaemia	Secondary Polycythaemias		
Some Collagen diseases	**D. WHITE CELLS**	ii Membrane	{ Surface Antibodies Heinz bodies
Hyperfibrinogenaemia.	Leukaemia		
		iii Contents	{ Abnormal haemoglobin Storage lesions Enzyme abnormalities Parasites

E. MIXED OR UNKNOWN AETIOLOGY

i Acute changes	{ Cerebrovascular accidents Myocardial infarction Following surgery
ii Chronic changes	{ Ischaemia of the legs Diabetes mellitus (retinopathy) Oral contraceptives Raynaud's phenomenon Myocardial ischaemia Cerebral ischaemia Essential hypertension

Fig. 2.   Classification of conditions associated with hemorheological abnormalities.

demonstrated can be partly classified on an etiological basis
(Fig. 2), whether the hemorheological defect is due to an abnormal
plasma protein concentration (Group A), a high hematocrit (Group B),
or decreased red cell deformability (Group C). This latter group
can be subdivided on the basis of whether the decreased red cell
deformability is due to the change of the red cell, the membrane,
or a cytoplasmic abnormality. We are increasingly recognizing
the rheological implications of the white cell (Group D). For the
most part, these first four groups represent classical hematological
diseases, and are collectively sometimes called the primary hemorheological diseases. More interesting are the common circulatory
diseases, such as intermittent claudication (6,7), myocardial
ischemia (8), Raynaud's phenomena (9), or venous thrombosis (10),
which form the last group. These are sometimes called secondary
hemorheological diseases, where the cause of the rheological abnormality is often multifactorial and not necessarily present in
all cases with these conditions.

VENOUS THROMBOSIS

The most direct evidence linking hemorheological abnormalities

to thrombosis is on the venous side where an increase in whole
blood viscosity, by slowing blood flow viscosity, may play a primary
etiological role. The lowest shear rates in the circulation are on
the venous side, and therefore, measurements of blood viscosity at
low shear rates are particularly relevant. The effect of fibrinogen
is particularly marked at low shear rates. As blood flow in these
regions is frequently intermittent, the yield stress of the blood
may also be significant, as this is the minimum force necessary to
begin movement in a static column of blood. In most clinical
conditions associated with deep venous thrombosis, such as surgery,
myocardial infarction, dehydration, shock, polycythemia, and
leukemia, the whole blood viscosity is also abnormally raised.
Apart from such circumstantial evidence, there is also some direct
evidence from prospective studies linking a high pre-operative
whole blood viscosity to the development of thrombosis following
surgery (10,11). In a study of 1,388 operations, during the course
of which some patients' viscosity was artificially lowered by hemo-
dilution, Duruble showed that almost all the clinically apparent
pulmonary emboli occurred in the patients whose hematocrit and
viscosity was not lowered (12). In the rather special case of
retinal vein thrombosis, a correlation between the severity of the
ischemia and whole blood viscosity as well as many of its determinants has also been demonstrated (13,14).

ARTERIAL THROMBOSIS

On the arterial side of the circulation, hemorheologically
related thrombosis is probably more important, but also more difficult to evaluate. The possible connections can be considered
under five headings.

1. Epidemiological Evidence

There is considerable body of evidence, both in relation to
hematocrit and plasma fibrinogen, which shows that they are primary
risk factors for the development of a variety of circulatory abnormalities associated with narrowing of arteries by atheroma and
thrombosis. The most plausible explanation is the critical influence
of both hematocrit and fibrinogen on blood viscosity. The classic
example in relation to hemoglobin concentration is the Framingham
Study which showed a close correlation between initial hemoglobin
concentration and the subsequent development of cerebral ischemia

and hypertension (15). A number of studies have also shown that a high hemoglobin concentration, even within a normal range, is a significant risk factor for the development of myocardial ischemia (16), cerebral ischemia (17), and peripheral arterial occlusion (18,19). The whole question of what is the optimal hemoglobin concentration in patients with arterial narrowing is undergoing radical rethinking (20).

As regards fibrinogen concentration and blood viscosity, high levels have been shown to be significantly associated with increased mortality in diabetics (21), intermittent claudicants (22,23), cerebral ischemia (24), and myocardial ischemia (25). In all these epidemiological or longitudinal studies the increased red cell or fibrinogen concentration, both invariably associated with hyper-viscosity, have been shown to cause an increase of hard circulatory endpoints at least in part due to thrombosis.

2. Initiation of Thrombogenesis

There are various possible mechanisms linking thrombogenesis directly to hemorheological and hemodynamic events. Born suggested that platelet aggregation and thrombosis may be initiated by the release of ADP from red cells. This would tend to occur at nar-rowing due to atheromatous plaques or at the site of hemorrhage into these plaques, where increased hemodynamic stresses may cause hemolysis or the release of ADP from intact red cells undergoing extreme deformation (26). The proposition that ADP released by red cells may play an important chemical role in the initiation of thrombi was already suggested more than two decades ago (27,28).

Turitto and his colleagues have long emphasized the role of the red cell in thrombus formation. They have shown that an increase in hematocrit has a marked effect not only on platelet adhesion to the subendothelium, but also on thrombus formation. This is partly a mechanical effect and partly some type of biochemical effect of the red cells, which appears to be shear-dependent, and therefore, possibly related to the mechanical properties of the red cell (29). The phenomenon is now generally accepted, the exact mechanisms are still in doubt (30). In an experimental model Rieger has also shown that high shear stresses alone can cause the release of intragranular constituents of platelets. But whether the mechanism is related to the red cells increasing the movement of platelets towards the wall (31), or the release of ADP from the

red cells due to their deformation or hemolysis, the effect is
related to the microrheology of the red cells. Recently, there
has been clinical direct evidence associating decreased red cell
filterability, and presumably deformability, with platelet adhesiveness. It is interesting that the hypothesis envisaging
the activation of platelets at areas of high shear stress initiating
arterial thrombosis is the opposite of the probable mechanism on
the venous side of the circulation where thrombosis tends to occur
at areas of low shear stress and shear rate. Blood of an abnormally
high viscosity would tend to favor both mechanisms; on the venous
side it would tend to decrease the shear rate or flow viscosity
for any given force, while on the arterial side it would increase
the stress on the red cells at the site of increased flow or
turbulence, for instance, at the site of irregularities or breaks
in the arterial lining.

3. Localization of Thrombi

Hemorheology may also play a part in the development and
localization of thrombi by its influence on the hemodynamic stresses
in arteries. The pattern of blood flow at the site of various
types of irregularities in the wall, stenosis or bifurcations have
been extensively studied and reviewed by Goldsmith and Karino (32).
The localization of atheromatous plaques, and probably also of
thrombi, to certain anatomical sites is most likely to be the
result of special local hemodynamic circumstances. This may take
the form of flow separation and vortices with stagnant centers (33),
or the effect of local high fluid stresses on the vessel wall.
All these phenomena are directly the result of the interaction
between the anatomical configuration of the vessels and the hemorheological properties of the blood flowing through them. The
exact definition of these forces and their effect in vivo is
extremely complicated and most of the theoretical work is based on
models. But, for instance, it has been shown in animals that the
endothelium is more rapidly replaced at the site of high wall
stresses and turbulence near the origin of branches (34,35). More
recently, endocytosis, prostacyclin production, as well as the
platelet reactivity of endothelial cell cultures have been shown
to be shear rate-dependent (36).

4. Clinical Evidence - Severity and Prognosis of Ischemic Disease

In many of the common diseases listed in Figure 2 the clinical
severity and progress of a patient's ischemia can be related to
the magnitude of the hemorheological abnormality. This has been
repeatedly shown in relation to intermittent claudication (6,22,23,
37), cerebral infarction (38-40), coronary disease (25,41-43),
diabetes (44,45), and many other less common conditions as well.
Rethrombosis following arterial reconstructive surgery is probably
the biggest single problem facing the surgeon treating peripheral
ischemic disease. In this area also, longitudinal studies have
shown that a raised hematocrit or fibrinogen is a severe risk
factor (46,47).

Perhaps one of the most fascinating new clinical findings in
the area of the hemorheological pathophysiology of acute ischemia
relates to the so-called coronary "spasm." Like diabetic micro-
angiopathy or essential hypertension, coronary spasm has often
been a negative diagnosis largely made by excluding other known
causes of myocardial ischemia. Recently, plasma viscosity and red
cell deformability have both been shown to be abnormal in patients
with documented myocardial ischemia and normal coronaries (48,49).
It is possible that the undoubted myocardial ischemia experienced
by some patients with apparently normal coronary arteries is not
due to "spasm," but to a hemorheological accident.

As before, the associations mentioned between the severity and
progress of a circulatory disease on one hand and a hemorheological
abnormality on the other, do not necessarily prove an association
between the latter and arterial thrombosis. However, it is generally accepted that the severity or the deterioration of circulatory
diseases is probably due to thrombosis (50) or progression of
atherosclerosis. Furthermore, one of the most popular theories of
atherogenesis is based on the incorporation of microthrombi into
the arterial wall.

5. The Ischemic-Hemorheological Vicious Cycle

A number of mechanisms linking thrombogenesis with hemorheological abnormalities have been considered. Lastly one should
consider the secondary changes in the rheological determinants
once ischemia has developed in a particular region. Local changes

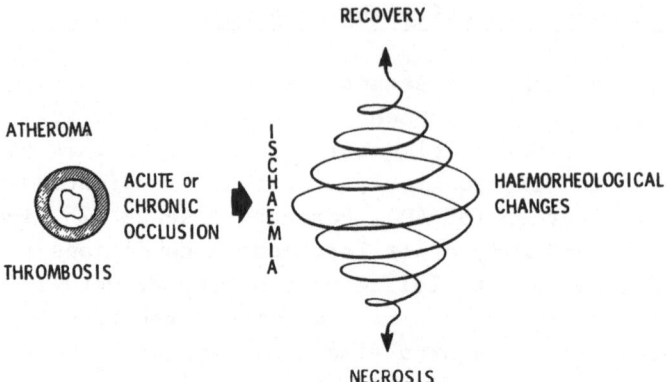

Fig. 3. Diagramatic representation of the
         vicious cycle hypothesis between
         tissue ischemia and hemorheological
         changes.

in ischemic tissue, such as hypoxia, hyperosmolarity, acidosis, and
the accumulation of metabolites, have all been shown to impair the
deformability of red cells and to increase blood viscosity (to
some extent these are part of the acute phase reaction). There may,
therefore, be set up a vicious cycle where ischemia decreases the
flow properties of the blood locally, which in turn further impair
the circulation increasing the severity and extent of the ischemia.
This is illustrated diagramatically in Figure 3. This hypothesis
would explain the strong prognostic significance of the early
hemorheological changes following acute tissue infarction. The
hemorheological changes may thus be both a consequence and a cause
of the ischemia. Further light has been thrown on the interaction
of ischemia and hemorheological changes by some very recent observations, suggesting that there are acute and reversible hemorheological changes which can be observed during intermittent
ischemic attacks such as exercise-induced angina. Most of this
pioneering work has been carried out in Professor Di Perri's
department (7,51). In the context of the more extreme ischemia or
myocardial infarction, the vicious cycle may spiral one way towards
necrosis or built-in physiological compensatory mechanisms may
help it spiral towards recovery. This balance may exist in both
chronic and acute ischemia, and it may be the logical point for
therapeutic intervention. A recent clinical demonstration of this
general theory has been the finding of an immediate decrease in
red cell filterability following myocardial infarction. Red cell
deformability often falls to less than half the normal value, with

recovery over the subsequent few days. The extent of the early fall in red cell deformability in individual cases was closely correlated to the subsequent clinical course of the patients in terms of pulmonary edema, cardiogenic shock, and mortality (52).

CONCLUSION

Neither the present nor our likely future understanding of hemorheology is going to provide the answer to the riddle of clinical thrombogenesis, any more than our increasing knowledge of clotting factors, hemostasis, platelets or prostaglandins has in the past. Hemorheological principles are, however, closely linked with all these other mechanisms and deserve equal scrutiny. This has not been the case in the past, and it is hoped that this brief presentation may have gone some of the way towards redressing the imbalance and encouraging further study of the role of hemorheology in thrombosis.

REFERENCES

1. J. Dormandy and H.L. Reid, Controlled defibrination in the treatment of peripheral vascular disease, Angiology, 29:80 (1978).
2. J.L. Wautier, M.P. Wautier, D. Pintigny, F. Galacteros, P. Passa, and J.P. Caen, Erythrocyte adhesion to endothelium in diabetes mellitus and sickle cell disease: its relation to vascular complications, Microcirculation Meeting, Jerusalem (1982).
3. S. Chien, Biophysical behaviour of red cells in suspencions, in: "The Red Blood Cell," Vol. II, Academic Press, London (1975).
4. J. Dormandy, Blood viscosity and red cell deformability, in: "Methods in Angiology," M. Verstraete, ed., Martinus Nijhoff, London (1980).
5. Proceedings of the London Workshop on Red Cell Deformability Martinus Nijhoff, London (in press) (1982).
6. H.L. Reid, J.A. Dormandy, A.J. Barnes, P.J. Lock, and T.L. Dormandy, Abnormal red cell deformability in patients with ischemia of the leg, Lancet, ii:666 (1976).
7. T. Di Perri, S. Forconi, M. Guerrini, S. Pecchi, and R. Cappelli, Modificazioni emoreologiche nelle arteriopatie, Giorn. Geront., XXIX:487 (1981).

8.   V.T. Turitto and H.J. Weiss, Platelet and red cell involvement
     in mural thrombogenesis. Presentation at New York Academy
     of Sciences Meeting (1982).

9.   A.J. Dodds, M.J.G. O'Reilly, C.J.P. Yates, L.T. Cotton, P.T.
     Flute, and J.A. Dormandy, Haemorheological response to
     plasma exchange in Raynaud's syndrome, Br. Med. J., 1186
     (1979).

10. J.A. Dormandy and J.B. Edelman, High blood viscosity: an
     aetiological factor in venous thrombosis, Br. J. Surg.,
     60:187-190 (1973).

11. W.V. Humpreys, A. Walker, and D. Charlesworth, Altered viscosity
     and yield stress in patients with abdominal malignancy,
     relationship to deep vein thrombosis, Br. J. Surg., 54:134
     (1976).

12. M. Duruble, Incidence of haemodilution on thromboembolic
     complications, Phlebologie, 34(1):13-19 (1981).

13. C.P. Ring, T.C. Pearson, M.D. Sanders, and G. Wetherley-Mein,
     Viscosity and retinal vein thrombosis, Br. J. Opthamol.,
     60:397 (1976).

14. G.D.O. Lowe, G. Trope, B.M. Mcardle, and J.T. Douglas, Abnormal
     blood viscosity and haemostasis in chronic retinal vein
     thrombosis, Presented at VII International Congress on
     Thrombosis, October (1982).

15. W.B. Kannel, T. Gordon, P.A. Wolf, and P. McNamara, Haemoglobin
     and the risk of cerebral infarction. The Framingham Study,
     Stroke, 3:409-420 (1972).

16. L.E. Bottiger and L.A. Carlson, Early phases of coronary heart
     disease, The Stockholm Prospective Study 2, Skandia Int.
     Symposia, Nordinska Bokhandelns Forlag, Stockholm, pp. 158-
     181 (1973).

17. H. Toghi, H. Yamanouchi, M. Murakami, and M. Kameyama, Impor-
     tance of the haematocrit on a risk factor: cerebral
     infarction, Stroke, 9:369-374 (1978).

18. M. Stafanini, J.V. Urbas, and J.E. Urbas, Gainsböck's syndrome:
     its hematologic, biochemical, and hormonal parameters,
     Angiology, 45:198 (1978).

19. P.S. Burge, W.S. Johnson, and T.A.J. Prankerd, Morbidity and
     mortality in pseudopolycythaemia, Lancet, i:1266-1269
     (1975).

20. J.A. Dormandy, C.J.P. Yates, and A. Berent, Is haemodilution
     a good thing? La Ricerca Clin. Lab., 11(Suppl. 1):173-177
     (1981).

21. A. Hart, H. Cohen, and J.M. Thorp, Lipoprotein and fibrinogen studies in diabetes, Postgrad. Med. J., June(Suppl.):435 (1971).

22. J.A. Dormandy, E. Hoare, J. Colley, D.E. Arrowsmith, and T.L. Dormandy, Clinical, haemodynamic, rheological, and biochemical findings in 126 patients with intermittent claudication, Br. Med. J., 4:581-583 (1973).

23. J.A. Dormandy, E. Hoare, A.H. Khattab, D.E. Arrowsmith, and T.L. Dormandy, Prognostic significance of rheological and biochemical findings in patients with intermittent claudication, Br. Med. J., 4:681-683 (1973).

24. L.O. Pilgeram, Abnormalities in clotting and thrombolysis as a risk factor for stroke, Thromb. Diath. Haemor., 31:245 (1974).

25. J.W. Meade, R. Chakrabarti, A.P. Haines, W.R.S. North, Y. Stirling, and S.G. Thompson, Haemostatic function and cardiovascular death: early results of a prospective study, Lancet, i:1050-1054 (1980).

26. G.V.R. Born, Arterial thrombosis and its prevention, Proceedings of 8th World Congress Cardiology, Hayase & S. Murao, eds., Tokyo, Elsevier, Amsterdam (1978).

27. A.J. Hellem, Release of ADP in the initiation of thrombi, Scand. J. Clin. Lab. Invest., (Suppl. 51):1 (1960).

28. A. Gaarder, J. Jonsen, S. Laland, A. Hellem and P.A Owren, Release of ADP by red cells in the initiation of thrombi, Nature, 192:531 (1961).

29. V.T. Turitto and H.J. Weiss, Red blood cells: their dual role in thrombus formation, Science, 207:541-543 (1980).

30. J.A. Davies, E. Essien, J.P. Caznave, R.L. Kinlough-Rathbone, M. Gent, and J.F. Mustard, The influence of red blood cells on the effects of aspirin and sulphin pyrazine on platelet adherence to damaged rabbit aorta, Br. J. Haematol., 42: 283-291 (1979).

31. V.T. Turitto and H.R. Baumgartner, Platelet interaction with subendothelium in a perfusion system: role of red blood cell, Microvasc. Res., 9:335-344 (1975).

32. H.L. Goldsmith and T. Karino, "Quantitative Cardiovascular Studies," N.H.C. Hwang, D.R. Gross, and D.J. Patel, eds., University Press, Baltimore (1979).

33. C.M. Rodkiewicz, Localisation of early atherosclerotic lesions in the aortic arch in the light of fluid mechanisms, J. Biomechan., 8:149-156 (1975).

34. D.L. Fry, Acute vascular endothelial changes associated with
     increased blood velocity gradients, Circ. Res., 22:165-197
     (1968).

35. H. Payling-Wright and G.V.R. Born, Possible effect of blood
     flow on the turnover rate of vascular endothelial cells,
     in: "Theoretical and Clinical Haemorheology," H. Hartert
     and A.L. Copley, eds., Elsevier, Amsterdam (1971).

36. M.A. Gimbrone, Jr., C.F. Dewey, Jr., P.F. Davies, and S.R.
     Bussolari, Hemodynamic shear stress and endothelial cell
     function: in vitro studies, Thromb. Haemostas., 478:153
     (1981).

37. B. Angelkort, N. Maurin, and K. Boateng, Influence de la
     pentoxifylline sur la deformabilite des hermaties dans les
     arteriopathies obliterantes peripheriques, Curr. Med. Res.
     Opin., 6:255-258 (1979).

38. E.O. Ott, H. Lechner, and A. Aranibar, High blood viscosity
     syndrome in cerebral infarction, Stroke, 5:330-333 (1974).

39. M.R. Boisseau, M.F. Lorient, J.M. Orgogozo, C. Doutremepuich,
     P. Martin, and H. Bricaud, The importance of red blood
     cell deformability in cerebrovascular disorders, in:
     "Disorders of Blood Flow. New Therapeutic Aspects,"
     R.V. Manrique and R. Muller, eds., Excerpta Medica,
     pp. 80-83 (1981).

40. D.A. Ionescu, M. Ghitescu, I. Marcu, and A. Xenakis, Erythrocyte
     rheology in acute cerebral thrombosis, effects of ABO
     blood groups, Blut, 30:(5)351-357 (1979).

41. G.D.O. Lowe, M.M. Drummond, A.R. Lorimer, I. Hutton, C.D.
     Forbes, C.R.M. Prentice, and J.C. Barbanel, Relation
     between extent of coronary artery disease and blood
     viscosity, Br. Med. J., 58:54 (1980).

42. E. Volger, K. Ostner, J. Klein, and A. Wirtzfeld, Changes in
     red cell aggregation and deformability after acute
     myocardial infarction, Proc. of Europ. Soc. of Microcirc.
     Microvasc. Res., 17:153 (1979).

43. T. Di Perri, S. Forconi, M. Guerrini, C. Rossi, and D. Pier-
     galli, Blood viscosity in the different stages of ichaemic
     heart disease, angina pectoris, and myocardial infarction,
     Excerpta Medica, I.C.S., 491:349-353 (1979).

44. G.D.O. Lowe, J.M. Lowe, and M.M. Drummond, Blood viscosity in
     young male diabetics with and without retinopathy,
     Diabetologia, 18(5):359-363 (1980).

45. J. Barnes, Hyperviscosity: a reversible factor in diabetic
     microcirculatory disease, Br. Med. J., 54:95 (1980).

46. J. Bouhoutsos, T. Morris, D. Chavatzas, and P. Martin, The
    influence of haemoglobin and platelet levels on the results
    of arterial surgery, Br. J. Surg., 34:287 (1979).
47. P.L. Harris, D.R. Harvey, and B.P. Bliss, The importance of
    plasma lipid, glucose, insulin, and fibrinogen in femoro-
    popliteal surgery, Br. J. Surg., 65:197-200 (1978).
48. D. Frimark, A. Slonim, and N. Cristal, Rheological factors
    among patients with ischemic heart disease and normal
    coronary arterias, Unpublished data.
49. R. Johnsson, M.H. Frick, M. Valle, P. Vuopio, and L. Tarssanen,
    Red cell flexibility and oxygen affinity in patients with
    angina pectoris and normal coronary arterial, Ann. Clin.
    Res., 13:77-80 (1981).
50. W.F.M. Fulton and D.J. Sumner, Radiofibrinogen in a study of
    fatal myocardial infarction, Presented at VII Int. Conf.
    of Thrombosis and Haemostasis, London, Thromb. Haemostas.,
    42:38 (1979).
51. S. Forconi, M. Guerrini, P. Ravelli, C. Rossi, G. Ferrozzi,
    S. Pecchi, and G. Biasi, Arterial and venous blood viscosity
    in ischaemic lower limbs of peripheral obliterative
    arterial disease patients, J. Cardiovasc. Surg., 20(4):
    379-384 (1979).
52. A.J. Dodds, P.P. Matthews, M.J. Bailey, P.T. Flute, and J.A.
    Dormandy, Changes in red cell deformability and other
    haemorheological variables after myocardial infarction,
    Br. Heart J., 44:508 (1980).

BLOOD CELLS AND THROMBOSIS: THE RHEOLOGY AND

ACTIVATION OF THE COAGULATIVE PROCESS

Tullio Di Perri

Istituto di Patologia Special Medica e
Metodologia Clinica dell'Università di Siena
Siena, Italy

INTRODUCTION

An approach to the clinical relevance of hemorheology in terms
of circulatory physiology and the relationship with either pro-
or pre-thrombotic conditions can begin from our knowledge of blood
flow regulation.

Blood flow can be measured rather easily in man. Dilutional
methods are used to estimate circulatory output particularly at
the onset of greater vessels, such as aorta or pulmonary arteries.
These methods are also used to measure cerebral or coronary
circulation, but their usefulness for pathophysiological and
clinical studies appears limited because they are stronly invasive
and give only instantaneous information. The continuous registration
of flow rate is, on the contrary, measurable in limbs by a non-
invasive technique such as strain gauge plethysmography based on
venous occclusion method. Our knowledge of peripheral blood circulation has been continuously enriched by such studies, but the data
refer to whole circulating blood without any indication of the
amount flowing in greater vessels or in microcirculation. In
mechanical terms, blood flow rate is dependent on several known
factors, theoretically assembled in Womersley's equation, derived
from the first Poiseuille law. In this synthetic expression, three
different forces can be analytically identified: the driving force
expressed by the pressure difference recorded at the two ends of
the vascular segment, the extrinsic resistence expressed mainly

by the fourth power of the vessel radius, and the intrinsic
resistance expressed by the viscosity of blood. The physiological
approach to blood circulation must take into consideration the
continuous interplay of these three factors, regulated to obtain
the requested flow rate and distribution. We must also take into
consideration that the driving force is not continuous, but
pulsable, and that vessel geometry is variable. Each determining
factor of blood flow is either measured or calculated separately
to permit the final calculation of the flow rate (24). The latter
is commonly referred to the time when each datum is recorded.
Moreover, in in vivo studies the forces opposing the progression
of flow, the so-called vascular resistances, are calculated together
by the circulatory output and pressure values, without discriminating
between extrinsic vascular resistance and intrinsic blood resistance.
The latter force is due to the dynamic interplay of several factors
in blood that together express the so-called blood viscosity, the
intrinsic friction opposing the tissue break induced by external
forces (3,9,17,26). Blood viscosity cannot be measured in vivo.
For several years, blood viscosity was measured in vitro ex vivo
giving a quantitative picture of the level of this physical property
of the blood. Studies of large populations conducted in many
laboratories throughout the world have collected useful data which
confirm the shear rate dependence of blood, as all typical non-
Newtonian fluid. Several values obtained from epidemiological and
naturalistic observations were employed to assess the so-called
"normal control" value referred to human subjects without any evident
physiological disorders and with a normal blood composition. The
normal control value was always related to shear rate, and was
statistically calculated in terms of mean, standard deviation,
sensibility and specificity of the method (3).

Blood viscosity determination must be integrated with the
assay of all known factors, such as hematocrit, plasma protein and
fibrinogen concentration, leukocyte and platelet count, aggregability
and deformability of red cells owing to their internal viscosity
state. The latter property was studied by filtrating either blood
or washed red cells in a closed system with a fixed negative pressure through a five micron nucleopore membrane evaluating the time
of filtration (2,17). All these data recorded ex vivo give a
code for interpreting in analytical terms the role played by
viscosity in the physiology of circulation, or, more realistically,
the viscosity changes associated with normal and abnormal circulatory conditions, which can be induced by experimental means, or

which can be observed in the course of spontaneous disease.

THE HYPERVISCOSITY SYNDROME: A PROTHROMBOTIC STATE

Presently, there is compelling evidence that increased blood
viscosity is associated with impaired microcirculation leading to
stasis and thrombosis (12,14). The so-called primary hyperviscosity
syndrome is considered as a consequence of diseases showing either
persistent increase of plasma protein concentration, such as in
macroglobulinemia and myelomatous disease, or a persistent increase
of red cell mass, such as in polycythemia, or an increase of internal
viscosity of erythrocyte, such as in some hemolytic disorders.
The three groups of diseases present a high risk to the development
of hyperviscosity syndrome with microcirculatory thrombosis due to
a common final mechanism, the slow down of circulatory rate, and
the impairment of blood vessel wall physiological interplay. It
is worthwhile underlining that even if the final mechanism leading
to microcirculatory obstruction is the same in the three types of
diseases, each disease shows a specific pathway which induces the
rheological disorder either by increasing plasmatic viscosity,
or by increasing hematocrit values, or through the modification of
the internal physico-chemical state of ertythrocyte. The three
pathways lead to a common final pathophysiological consequence,
the increase of viscosity of circulating blood and to a unique
clinical consequence, the microthrombosis particularly localized
in the central nervous system. It is generally accepted that
microcirculatory changes depend on increased viscosity of the blood
in the above mentioned disease. In such situations, the analytical
dissection of the viscosity skeleton is absolutely necessary to
confirm the presumptive role of the specific pathway working to
impair blood circulation in the smallest vessels (14).

Many clinical disorders, not apparently related to biochemical
changes of plasma or to cellular blood concentrations, are often
associated with inconstant increase in blood viscosity. Horizontal
and vertical clinical investigations registered that the common
link between the patients showing such finding was the presence
of an obstructing vascular disease in ischemic phase with presumptive
microcirculatory and tissutal derangement (2,3,5,6,8,13,17,25).
Actual evidence suggests that the rheological disorder is a consequence of impaired blood flow with ischemia closely dependent
on it (5,12,16,17). Consequently, it can be considered as a sign

of circulatory and metabolic trouble. The question arises whether
this ischemia-dependent hyperviscosity can work as an autoalimen-
tation mechanism, recruiting ischemia itself and closing a pathogenetic vicious cycle (9,25). One of the most exciting problems,
both from the speculative and the clinical point of view, is blood
hyperviscosity in diabetic patients. The association with micro-
angiopathy appears to be confirmed, and the vertical studies suggest that hyperviscosity could be considered as a risk factor,
with high predictive value for the microangiopathic fate of the
diabetic patient. The viscosity disorder appears to be mainly
due to intrinsic metabolic alterations of erythrocyte, but extrinsic
plasmatic viscosity factors also seem to play a role. A delayed
filtration rate of erythrocyte is the first and sometimes the only
sign of rheological impairment. Then the whole blood viscosity can
increase, but the study of a single viscosity factor shows that the
onset of hyperviscosity syndrome in the diabetic patients is associated with an increased rigidity of circulating cells and
particularly of red cells (3,6).

ROLE OF BLOOD CELLS IN HYPERVISCOSITY: ERYTHROCYTE, POLYMORPHO-NUCLEATE AND PLATELET

After clinical observations of hyperviscosity syndrome associated with leukemia and thrombocythosis, our attention is now
focused on the role of other cellular components of blood, leukocytes and platelets (3). It appears rational and in line with the
mechanic of fluids that the increase of the number of white cells
and/or of platelets should influence Poiseuille flow (24). After
the results registered in obstructive vascular patients during
spontaneous or induced ischemia, a different approach to the problem
of influence that leukocytes and platelets circulating in blood have
on its rheological profile was recently proposed by our laboratory.
In such patients, either with ischemic cardiac disease or with
peripheric obstructive arterial disease, the appearance of an
ischemic crisis, respectively angina pectoris or claudication,
was associated with a rapid decrease of blood filtration rate and
an increase of whole blood viscosity (5,15,18,20). The viscosity
disorder, rapidly appearing and disappearing, was clearly related
to the effect of ischemic process. The first interpretation, supported by regional artero-venous viscosity studies, was that such
hyperviscosity could be due to a slowdown of blood flow rate in
ischemic vascular region, including metabolic-flogistic reaction of

hypoperfused tissues and secondary local changes of blood viscosity
(21).   Later on, it was noted that even if this mechanism is
working, other pathways seem to be involved, since rheological
changes develop far from the ischemic area and independently from
the local circulatory alteration. The first naturalistic finding
was that the development of an ischemic reaction, whereever it may
be, either in the coronary tree or in peripheral arteries, induced
either spontaneously or by exercise, was associated with marked
increase of blood β-thromboglobulin concentration, while in control
subjects, even strenous exercise was not followed by such changes
(10).   The observation of contemporaneous modification of whole
blood rheology and particulary of blood filtrability on one hand,
and of activation of circulating platelets on the other hand, suggested the possibility of a common pathway leading to both biological
alterations, which are most likely to be dependent on some mechanisms
secondary to the ischemic process (10,16). In intravascular platelet
aggregation, recent observations showed that hemodynamic and biochemical interactions occur, since induction of platelet activation
may be induced by abnormal hemodynamic conditions with wall shear
forces leading to erythrocyte lysis and dismissal of cytoplasmatic
content. Red cell cytoplasm is rich in ADP which is a known
inducer of platelet aggregation. In vitro assays with platelet-
rich plasma incubated with hemolysate confirm this hypothesis. To
explain peripheral localization of arterial thrombosis, Born (1)
has for a long time proposed such a mechanism which operates in
vivo together with other chemical and mechanical-dependent induction.
However, in biological terms, we must admitt the existence of the
possibility of platelet activation due to mechanical and chemical
hemolysis. The finding of an in vivo activation following ischemia
has suggested to us a variant of Born's hypothesis, which considers
erythrocyte deformability as the dependent variable and the released
content of platelet as the determinant variable. In vitro studies
in our laboratory showed that incubation of whole blood with the
supernatant phase of platelet-rich plasma (PRP) was associated with
an increase of filtration time, only if PRP was pretreated with an
inducer of aggregation, such as ADP. The experimental design is
rather complex, since PRP was withdrawn both from a normal control
subject and from an obstructive vascular patient, either in a
resting state or after ischemia induced by exercise. The collected
findings showed that filtration rate of erythrocyte (in whole blood)
was influenced by the supernatant phase of aggregated platelets in
progressive escalation, since the control platelet was the least

active in slowing the filtration rate, while the vascular patient
platelet was the most active, particularly after strenous exercise.
Statistical analysis stressed these differences.

Our interpretation was that, during ischemia, in vivo activation
of platelets occurs and influences blood rheology and especially
microrheology, since the so-called filterability clearly appears
worsened (10,16). Cellular components of blood, mainly red cells,
play an evident role in microrheology, either in terms of cell
concentration or in terms of membrane fluidity. Membrane fluidity
is a multifactorial-dependent variable influenced by internal
cellular metabolic pathways and external surface interrelationships
with an elevated number of endogenous and exogenous chemical
substances. Our findings seem to support the hypothesis that a
cellular component released by activated platelet modified blood
filterability, perhaps by changing membrane fluidity of red cells
with measurable change in blood rheology. Preliminary results on
washed red cell are in keeping with this interpretation. This
mechanism must be considered reversible in a short time, strongly
dependent on ischemic process, and mediated by biochemical pathway,
which goes from platelet cytoplasm to erythrocyte membrane. In
this sense, the mechanism of interrelationship proposed by Born,
suggesting a direct action of hemolysis on platelet activation, can
be broadened to include the possibility of an inverse direction,
which suggests a direct action of platelet content of erythrocyte
membrane fluidity.

Recent findings in our laboratory showed that pretreatment with
ASA or ticlopidine in obstructive vascular patients cancels the
exercise-dependent activity of platelet on blood filtration rate.
These data seem to agree with the hypothesis that platelet activation
can influence blood rheology by releasing substances which are
active on other blood components (10,16).

The most recent evidence of cellular blood functional influence
on rheological balance concerns polymorphonuclear leukocytes. The
association of a hyperviscosity state with microvascular thrombosis
in leukemic disease with a marked increase in circulating leukocytes
was observed, and the role of the increased number of leukocytes
was shown as a rheological variable capable of inducing an imbalance
between cells and plasma in circulating blood, in analogy with the
mechanism considered to be operating in polycythemia. A new
hypothesis was at the basis of experimental studies conducted in

our laboratory. The possibility of an ischemic-induced polymorpho-
nucleate activation was the first hypothesis. Very recent findings
have also shown that in obstructive vascular patients the functional
behavior of PMN changed after ischemic exercise. Several PMN
functions were studied: the superoxide production, enzyme release
and aggregation after specific stimulation (FLMP or PLM). After
having undergone exercise, it was shown that in vascular patients
an increase in leukocytic superoxide release in vitro appeared to
indicate a certain level of in vivo change of the leukocytic state,
which can be considered a sign of functional activation. These
patients at the same time showed an increase in β-thromboglobulin
level and a decrease in blood filtration. These findings were
integrated with the study of blood filtration after incubation with
supernatant phase of in vivo PMN activated by specific inducers.

In this study, it was shown that such incubation influences
blood filterability, thus suggesting that PMN cellular-released
substances find target molecular structures that are involved in
microrheological balance. In the actual hypothesis, the red cell
membrane appears to be the most rational target for this kind of
influence. Therefore, the primary concept of the mechanical red
cell properties, as a main factor of microrheological regulation,
must be broadened to take into consideration the biochemical
influence on mechanical behavior of erythrocyte coming from the
blood cells. As previously stated, the role played by platelets
and PMN does not, therefore, depend only on their number, but on
their functional state, since many physiological conditions, such
as ischemia, can rapidly induce activation of their biological
properties (26). The trigger mechanism of secondary hyperviscosity
is, generally speaking, the circulatory and metabolic disorders
provoked by ischemia (8).

The molecular approach to its definition has yet to be
investigated, and we do not know at which level of the vascular
and cellular cascade induced by metabolic insufficiency the
rheological platelet and PMN changes appear, nor the timing of
these phenomena, nor the anatomic site from which they start.
Since vascular dilation, secondary to ischemia, mainly occurs in
microvessels, like arterioles, metabolic changes are more likely
to develop in the exchange tree, the capillary bed, and the surrounding tissue. In our laboratory, the metabolic and rheological
studies on obstructive arterial diseased leg, evaluating O_2, CO_2,

lactate, pH arteriovenous difference, and venous blood viscosity
and filterability after exercise selectively made only on the
diseased leg, showed that soon after the onset of exercise A.V.
difference of PO2 decreased while that of pCO2 increased with a
very marked increase of A.V. difference of lactate and a fall in
venous pH. At the same time, femoral venous viscosity increased
and blood filterability decreased to a considerable amount. The
comparison with findings reported in the control subject, with the
same experimental model, showed a marked difference, since even
strenous exercise is followed only by very small and rapidly disappearing metabolic and rheological changes, thus underlining that
the impairment of active hyperemia could be considered the trigger
of the glycolytic anaerobic shift which appears associated with
rheological abnormalities (23). Without discussing the dependence
of the observed phenomena, we can propose the hypothesis that both
metabolic and rheological changes originate from the same functional
environment, the microcirculatory tree. In this sense, rheological
modification observed in obstructive vascular patients can be
interpreted as an indirect sign of metabolic tissue insufficiency,
and in vertical studies, it may be helpful in evaluating the course
of the disease in the single patient and the danger of its occlusive fate. In this sense, the clinical rheological study appears to be of interest in the evaluation of the so-called thrombotic
evolution of the obstructive disease (4,16).

REFERENCES

1.   G.V.R. Born, Haemodynamic and biochemical interaction in intra-
        vascular platelet aggregation, in: "Blood Cells and Ves-
        sel Walls," Ciba Foundation Symposium 71, 61 (1980).
2.   S. Chien, The present study of blood rheology, in: "Hemo-
        dilution: Theoretical Basis and Clinical Application,"
        K. Messner and H. Schmidt-Schönbein, eds., Karger, New York
        (1972).
3.   L. Dintenfass, Blood microrheology-viscosity factors in blood
        flow, in: "Ischaemia and Thrombosis," Butterworths, London
        (1976).
4.   T. Di Perri, S. Forconi, M. Guerrini, C. Rossi, M.S. Verzuri,
        M. Materazzi, and D. Pieragalli, Die Viskosität des blutes
        bei vaskülaren erkrakungen, From Medici, 37 (1983).
5.   T. Di Perri, S. Forconi, M. Guerrini, C. Rosssi, and S. Pecchi,
        Modificazioni della viscosità ematica sistemica in soggetti

con vasculopatie croniche distrettuali durante ischemia
spontanea o provocata, Boll. Sic., 22:1138 (1977).

6. T. Di Perri, Basi fisiopatologiche del trattamento farmacologico
    delle vasculopatie, Farmaci, 1/3:19 (1977).

7. M. Guerrini, A. Acciavatti, M. Materazzi, C. Rossi, C. Del
    Bigo, S. Forconi, and T. Di Perri, Protective effects of
    buflomedil against exercise-induced reductions in regional
    erythrocyte deformability of patients with peripheral
    arterial disease, J. Int. Med. Res., 10:387 (1982).

8. T. Di Perri, S. Forconi, M. Guerrini, C. Rossi, M.S. Verzuri,
    M. Materazzi, and D. Pieragalli, Pathophysiological
    meaning of blood viscosity changes in vasculopathic
    patients, Quaderni della Coagulazione, S. Coccheri, ed.,
    Periodici Baldacci di Informazione Medica, Bologna, 112
    (1978).

9. T. Di Perri, Rheological factors in circulatory disorders,
    Angiology, 30:480 (1979).

10. T. Di Perri, A. Vittoria, M. Guerrini, D. Pieragalli, C. Del
    Bigo, G. Martelli, M. Franchi, and P. Blardi, Action of
    ASA on hemorheological changes of ischemic patient. ASA
    International Symposium on Present State of Acetylsalicylic
    Acid in Research and Therapeutic Application, G. Segre
    and H. Ohnmeiss, eds., O.I.C. Medical Press, Firenze, 81
    (1982).

11. T. Di Perri, S. Forconi, M. Guerrini, C. Rossi, and D. Piera-
    galli, Blood viscosity in the different stages of ichaemic
    heart disease, angina pectoris, and myocardial infarction,
    Excerpta Medica, International Congress Series, 491:349
    (1979).

12. T. Di Perri, Fisiopatologia generale dell'ischemia, in:
    "L'Ischemia," T. Di Perri, B. Fischetti, S. Forconi, E.
    Malan, and F. Pratesi, eds., Società Editrice Universo,
    Roma (1980).

13. T. Di Perri, S. Forconi, M. Guerrini, S. Pecchi, D. Pieragalli,
    R. Cappelli, and A. Acciavatti, Influence of non-selective
    and selective beta adrenoceptor blockade on isoxsuprine-
    dependent hemodynamic and rheological changes, Angiology,
    32:257 (1981).

14. T. Di Perri, Aspetti fisiopatologici e clinici delle sindromi
    da iperviscosità ematica, La Ricerca Clin. Lab. II,
    Suppl. I, 73 (1981).

15. T. Di Perri, S. Forconi, M. Guerrini, S. Pecchi, and R.

Cappelli, Modificazioni emoreologiche nelle arteriopatie, Giornale di Gerontologia, 29:487 (1981).

16. T. Di Perri, M. Guerrini, A. Vittoria, D. Pieragalli, M. Franchi, and G. Martelli, Emoreologia e trombosi, in: "Recenti Acquisizioni in Tema di Terapia Antipiastrinica," Masson Italia Editori, Roma (1982).

17. J.A. Dormandy, Clinical significance of blood viscosity, Ann. R. Coll. Surg. Engl., 47:211 (1970).

18. S. Forconi, M. Guerrini, D. Agnusdei, F. Laghi Pasini, and T. Di Perri, Abnormal blood viscosity in Raynaud's phenomenon, Lancet, ii:586 (1976).

19. S. Forconi, M. Guerrini, C. Rossi, S. Pecchi, and T. Di Perri, Modificazioni della viscosità ematica sistemica durante claudicatio intermittens provocata nella arteriopatia obliterante periferica, Boll. Sic., 22:441 (1977).

20. S. Forconi, M. Guerrini, C. Rossi, S. Pecchi, and T. Di Perri, Modificazioni della viscosità ematica sistemica durante prova da sforzo nell'angina pectoris, Boll. Sic., 22:435 (1977).

21. S. Forconi, G. Biasi, M. Guerrini, P. Ravelli, C. Rossi, G. Ferrozzi, and S. Pecchi, Arterial and venous blood viscosity in ischemic lower limbs of peripheral obliterative arterial disease patients, J. Cardiovas. Surg., 20:379 (1979).

22. S. Forconi, M. Guerrini, D. Pieragalli, and A. Acciavatti, Viscosimetria ematica: approccio metodologico, La Ricerca Clin. Lab. II, Suppl. II, 135 (1981).

23. M. Guerrini, A. Acciavatti, M. Materazzi, C. Rossi, C. Del Bigo, S. Forconi, and T. Di Perri, Protective effects of buflomedil against exercise-induced reductions in regional erythrocyte deformability of patients with peripheral arterial disease, J. Int. Med. Res., 10:387 (1982).

24. D.A. McDonald, "Blood Flow in Arteries," Vol. I, Williams and Wilkins, Baltimore (1974).

25. H. Schmidt-Schönbein, Microrheology of erythrocytes, blood viscosity and the distribution of blood flow in the micro-circulation, in: "International Review of Physiology," Vol. II, A.C. Guyton and A.W. Cowley, eds., University Park Press, Baltimore (1976).

26. M.G. Tonneson, L. Smedly, A. Goins, and P.M. Henson, Interaction between neutrophils and vascular endothelial cells, Cologne Atherosclerosis Conference (Agent and Actions Suppl. II), 25 (1982).

PROCOAGULANT ACTIVITY OF PLATELETS, LEUKOCYTES, AND CANCER CELLS

POSSIBLE ROLE IN HEMOSTASIS AND THROMBOSIS

Nicola Semeraro

Università di Bari
Istituto di Patologia Generale
Piazza G. Cesare
Bari, Italy

INTRODUCTION

The sequences of reactions culminating in the formation of thrombin are classically known as the intrinsic and extrinsic pathways. It is generally accepted, mainly on the basis of in vitro experiments, that the initiation of the intrinsic pathway involves the contact of blood with appropriate surfaces within injured vessel wall, and that the diffusion of tissue substances (tissue factor or thromboplastin) in the blood stream represents the trigger mechanism for the extrinsic pathway. However, the question concerning how blood coagulation is triggered in vivo during physiological hemostasis or in pathological conditions associated with fibrin deposition is still unanswered.

Over the past decade, several investigators have shown that some normal or pathological cells may express different types of clot-promoting activities through which they can trigger and/or accelerate the coagulation sequence. The purpose of this presentation is to review present knowledge on the coagulant activities of platelets, leukocytes, and cancer cells and on their possible role in hemostasis and thrombosis.

PLATELET COAGULANT ACTIVITIES

A. Catalytic Activities

    It has long been known that platelets contribute to blood coagu-
lation essentially by providing a catalytic surface capable of accelerating two important reactions of the clotting cascade, namely
the activation of factor X by factor IX_a, and the conversion of
prothrombin to thrombin by factor X_a. This property, referred to as
platelet factor 3 (PF3), exists in a latent form in intact circulating platelets; it becomes available when platelets are 'activated'
during the adhesion/aggregation reaction and is associated with some
negatively charged phospholipids (mainly phosphatidylserine). In
unstimulated platelets, these phospholipids are located almost exclusively in the inside surface of platelet membrane. Maximal PF3
activity (prothrombin-converting activity) is observed in completely
lysed platelets or following activation by the combined action of
collagen plus thrombin. It has been suggested that in these circumstances active phospholipids are transferred from the inner to
the outer monolayer of the platelet membrane (transbilayer movement
of phospholipids or 'flip-flop') (1). Recent studies have shown that
when platelets are stimulated with agents capable of inducing the
'release reaction' such as thrombin or arachidonic acid, a specific,
high affinity receptor for factor X_a is exposed on their surface (2).
Once factor X_a is bound to the platelet receptor, the rate of
thrombin generation is markedly accelerated (2). Platelet receptor
for factor X_a is, at least in part, identical to activated factor V
(2,3). Factor V found on the platelet membrane may possibly be
activated to factor V_a by platelet activation reaction, which occurs
at the site of a vascular lesion and then binds to appropriate
structures (of unknown nature) of platelet membrane. A platelet
activator of factor V has been reported by Osterud et al. (4), who
found that platelets would express factor V_a activity after treatment
with collagen. These observations suggest that optimal conversion
of prothrombin to thrombin on the platelet surface would be ensured
when the catalytic activity of PF3 is further enhanced by platelet
factor V_a. The pathophysiological importance of these platelet
properties is underscored by the observation that some patients with
hemorrhagic disorders have a selective deficit of PF3 and/or of
structures necessary for binding factor V_a-X_a to the platelet
membrane (5).

B. Platelet Activities Triggering Blood Coagulation

Over the last years a number of platelet coagulant activities
distinct from PF3 and from platelet receptor for factor X_a have been
described by which platelets, at least in well defined experimental
conditions, may trigger blood coagulation. Walsh et al. (6,7) have
suggested that platelets may play a significant role in the activation of the intrinsic pathway of blood coagulation. They have
shown that platelets stimulated in vitro by low concentrations of
ADP or collagen promote factor XII activation in the presence of
kallikrein and high molecular weight kininogen, thus providing a
contact activating surface. In addition, normal platelets contain
measurable amounts of factor XI; when stimulated by collagen or
thrombin, they promote the activation of surface-bound factor XI
in the presence of kallikrein and high molecular weight kininogen,
but in the absence of factor XII. This observation raises the possibility of an alternative pathway which bypasses factor XII, thus
providing a possible explanation for the absence of hemostatic
defects in patients with factor XII deficiency or Hageman trait.
Although the question of the physiological relevance of these studies
which utilize artificial incubation conditions is difficult to ad-
dress, Walsh et al. (6,7) postulated a fascinating hypothesis linking
the events of primary hemostasis and blood coagulation. As platelets
adhere and aggregate to form a hemostatic plug, they provide a surface for assembling, activating, and binding coagulation proteins in
a sequential manner starting from 'contact activation' and culminating in the local generation of thrombin and fibrin. The protection by platelets of activated clotting factors from inactivation
by natural inhibitors may further contribute to localization of the
clotting process in a protective nidus provided by aggregating
platelets.

The pathophysiological relevance of platelet coagulant activities is supported by the observation that platelets from patients
without detectable plasma factor XI and without hemostatic disorders
contain measurable amounts of factor XI and, when stimulated by
collagen, behave normally in the activation of intrinsic coagulation
(8). In contrast, in some patients with hereditary giant platelet
syndrome of Bernard-Soulier suffering from severe bleeding tendency,
platelets lack factor XI and fail to activate intrinsic coagulation
when stimulated with collagen (8,9). It is concluded that platelet
membrane factor XI can substitute plasma factor XI in the coagulation
mechanism, and may be required for hemostasis even when plasma

factor XI is present. Further studies by Walsh et al. (10-12) suggest that platelet coagulant activities may also play a role in some
thrombotic processes. Indeed, a close correlation was found between
increased platelet coagulant activities and postoperative venous
thrombosis detected by radioactive fibrinogen uptake (10). Moreover,
a significant increase of platelet coagulant activities was described
in patients with transient cerebral ischemia and in patients with
myeloproliferative disorders and thromboembolic complications
(11,12).

An additional role of platelets in blood coagulation is suggested by the observation that, in well defined experimental conditions, platelets express a weak coagulant activity that apparently
is capable of activating coagulation factor X independently of the
intrinsic and extrinsic pathways (13). Direct activation of factor X
has been also observed by van Wijk et al. (14) using purified human
factor X and washed human platelets. Neither the mechanism nor the
relative importance of this platelet property is clear, but there is
evidence that it may be often markedly reduced in those patients with
myeloproliferative diseases who have a bleeding diathesis (15).

More recent experiments have shown that human platelets develop
a strong coagulant activity during simultaneous incubation with
endotoxin and leukocytes in plasma (16). This activity is at least
partially similar to that mentioned above (apparently capable of
directly activating factor X). This endotoxin-induced, leukocyte-
mediated platelet coagulant activity could contribute to blood clotting activation and subsequent fibrin deposition frequently associated with endotoxemia (in particular gram-negative sepsis) and other
immune inflammatory diseases.

LEUKOCYTE PROCOAGULANT ACTIVITY

Normal human and animal (rabbit, dog, mouse) peripheral blood
leukocytes generate a potent procoagulant activity (PCA), identified
as tissue factor, on exposure to bacterial endotoxin. They are,
therefore, capable of triggering blood coagulation through the extrinsic pathway (17). This activity is practically absent in freshly
isolated, unstimulated leukocytes. PCA generation is almost completely inhibited by cycloheximide and actinomycin D, suggesting that
it is due to 'ex novo' synthesis of apoprotein III. More recent
studies have shown that the capacity to produce PCA in response to
endotoxin is confined to the mononuclear cell fraction, whereas,

polymorphonuclear leukocytes are completely inactive.

Other stimulants different from endotoxin can induce tissue factor generation by mononuclear cells including antigens, mitogens, immune complexes, complement proteolytic products C5a and C3b, platelet and platelet membranes, adherence to various surfaces etc. (17). It is now firmly established that the monocyte is the cellular source of PCA (17-21) and that mononuclear phagocytes (macrophages) from diverse anatomical sites (peritoneal, splenic, hepatic) also share this property (22-25). However, conflicting views exist on the cellular events involved in the induction of monocyte/macrophage PCA. Some investigators reported that the monocyte procoagulant response to endotoxin or immune complexes was absolutely dependent on the presence of T-lymphocytes (21,26). Others found that the monocytes per se could generate PCA on exposure to certain stimuli (20,27,28) and that T-lyphocytes, at most, facilitated maximal expression of the monocyte response (17,27). We recently found that highly purified (< 99%) human macrophages obtained by prolonged in vitro culture of adherent monocytes as well as pure (<95%) populations of human peritoneal and milk macrophages are capable of producing PCA identical to tissue factor when stimulated with endotoxin (29). Since in all instances our macrophage preparations contained very few, if any, lymphocytes it is apparent that these cells are autonomous in their procoagulant response to endotoxin. Further support to this concept derives from recent observations that human monocytoid cell line U-937 can produce tissue factor in response to various stimulants without any need for cellular cooperation (30).

The exact pathophysiological significance of monocyte/macrophage PCA remains to be established. In particular, it is not yet known whether this activity is of any relevance for physiological hemostasis. Considering the nature of the stimulants that induce PCA generation by mononuclear phagocytes, it is likely that these cells may play an important role in the activation of intra- and/or extravascular coagulation and subsequent fibrin deposition associated with several pathological conditions including immune inflammatory diseases, some types of venous thrombosis, tumors, and atherosclerosis. Indeed, the presence of mononuclear cell infiltrates concomitant with fibrin deposition in the areas of tissue damage is a prominent feature of these pathological processes. Several lines of evidence derived from experimental models and from studies conducted in man would support this concept.

The essential role of mononuclear phagocyte PCA has been fairly
well established in endotoxin-induced disseminated intravascular
coagulation (DIC), at least in rabbits (reviewed in ref. 31).
First of all, endotoxin administered to rabbits induces the generation of PCA by circulation and resident (peritoneal) mononuclear
phagocytes in vivo (31,32). Endotoxin-induced DIC is prevented
when the animals are rendered leukopenic by cytotoxic drugs.
However, the infusion of normal leukocytes in leukopenic animals
restores susceptibility to endotoxin-induced coagulative changes.
Moreover, the infusion of 'activated' leukocytes (i.e., with available PCA) to normal animals causes an immediate DIC. In humans DIC
is a major complication of Gram-negative sepsis. Although direct
evidence is lacking, some would suggest that mononuclear phagocytes
contribute to blood clotting activation in this pathological condition. For instance, Komp and Donaldson (33) reported that patients
with Gram-negative sepsis complicating acute leukemia and with
severe leukopenia presented no signs of DIC. They postulated that
leukocytes are necessary for DIC to occur as a result of endotoxemia. In addition in patients with chronic lymphatic leukemia
or with chronic myeloid leukemia, who have a very low incidence
of thromboembolic manifestations and/or DIC despite the frequent
occurrence of severe infectious diseases, peripheral blood mononuclear cells have a marked defect in their ability to produce PCA
in response to endotoxin (34,35). These findings suggest that not
only a sufficient number of leukocytes (33) but also a normal
leukocyte function in terms of procoagulant response is required
for DIC to occur as a consequence of endotoxemia.

Another pathological condition in which mononuclear phagocyte
PCA might play an important role is postoperative venous thrombosis.
Lerner et al. (36), using a model of experimental venous thrombosis,
have shown that the thrombotic lesion is characterized by the presence of leukocytes adherent to the venous endothelium and that these
leukocytes develop a strong PCA. Miller et al. (37) observed increased production of monocyte PCA in a group of trauma patients
with splenectomy. The highest levels of PCA were seen in patients
who subsequently developed thromboembolic complications. There was
a significant correlation between increased PCA and accelerated
fibrinogen turnover.

Patients with malignant disease often experience thromboembolic
complications and/or DIC (38). Rickles et al. (39) have shown that

peripheral blood mononuclear cells from patients with advanced
carcinoma of the lung generate increased amounts of PCA when incubated in vitro with and without various stimulating agents. It
is remarkable that a striking correlation was found between monocyte
PCA and plasma levels of fibrinopeptide A. The latter is a well
known sensitive indicator of the level of ongoing fibrin formation. These investigators suggested that mononuclear phagocyte
PCA could contribute to blood clotting activation during the course
of malignant disease. More direct evidence that macrophages may
actively participate in fibrin deposition at the tumor site derives
from a recent study by Lorenzet et al. (40), who reported that, in
the rabbit V$_2$ carcinoma, macrophages harvested from the tumoral
mass (tumor-derived macrophages) express strong PCA immediately
after isolation without any stimulation.

Recently, Schwartz et al. (41), in the process of analyzing
the effects of lipoproteins on function of lymphoid cells, have
observed that physiological concentrations of isolated human plasma
lipoproteins have the capacity to enhance the expression of PCA of
human peripheral blood mononuclear cells in vitro. In particular,
very low density lipoproteins induce the generation of a typical
tissue factor activity (identical to that induced by the stimulants
mentioned above), whereas, intermediate density lipoproteins and
high density lipoproteins stimulate the production of a different
PCA which appears to be a direct prothrombin activator (prothrombin-
ase). In contrast, low density lipoproteins do not stimulate PCA,
but instead actively suppress the generation of the two procoagulants induced by the other lipoproteins. Reconstitution is not
stimulatory probably as a consequence of the suppressive effect of
low density lipoproteins. The in vivo implications of these interactions are presently unknown. Conceivably, lipoprotein-induced
PCA may be important in thromboembolic manifestations associated
with certain types of hyperlipoproteinemia and possibly in the
development of atherosclerotic lesions.

CANCER CELL PROCOAGULANT ACTIVITY

The involvement of the hemostatic system in malignant disease
has long been known. First, malignant diseases is often associated
with thromboembolic complications and/or DIC. On the other hand,
using histologic, immunologic or radioisotopic techniques, several
investigators have repeatedly observed the presence of fibrin within

and around tumors. The mechanisms responsible for activation of
blood coagulation in malignancy is extremely complex (reviewed in
Ref. 38). We have already discussed the possible role of mono-
cyte/macrophage PCA (see above). A number of studies have suggested
that cancer cells per se could play a major role by virtue of various
procoagulants capable of triggering blood coagulation. Pineo et
al. (42) and Gordon et al. (43) reported that extracts or secretes
from human and experimental tumors have a peculiar procoagulant activity which initiates blood clotting by activating directly coagulation factor X (independently from the classical intrinsic and extrinsic pathways). These findings have been confirmed and extended
by our group using intact cells from murine tumors instead of tissue
extracts (44,45). It was shown that cells from Ehrlich carcinoma
ascites, cells from JW sarcoma ascites, and cells from Lewis Lung
carcinoma, isolated either directly from the tumoral mass or in
culture, generate factor X$_a$ when incubated with a source of factor X
and calcium chloride. Gordon and Lewis (46) found a similar activity
in the culture medium of some malignant and transformed cells.
Recently, the procoagulant principle has been isolated and characterized (47). It is a single polypeptide chain with a molecular weight
of 68,000 daltons having the physico-chemical and enzymatic properties of a cysteine protease. This distinguished cancer procoagulant
from the other clotting enzymes (serine proteases). Direct factor X
activation by the unique cysteine protease does not appear to be the
only pathway of blood clotting initiated by cancer cells. Typical
tissue factor, activating factor X through factor VII (extrinsic
pathway), may occur in some malignant cells. In humans, the para-
digmatic example is the leukemic cell. The production of tissue
factor by circulating promyeloblasts in acute promyelocytic leukemia
has been regarded as the archetypal contribution of malignant cells
to hemostasis and thrombosis (48). Isolated leukemic promyelocytes
have potent clot-promoting activity found mainly in the granule
fraction. Its presence is consistent with the very high incidence
of DIC in patients with acute promyelocytic leukemia. Tissue factor-
like activity has also been described in severe experimental tumors
(38,49).

     In conclusion, there are at least two pathways of blood clotting
initiation by cancer cells (direct activation of factor X and extrinsic pathway) both requiring calcium ions. Although the exact
pathophysiological implications of cancer procoagulants activities
are not yet well known, recent studies suggest that not only could
they contribute to the activation of intravascular coagulation in

cancer patients, but they could also play an important role in the processes of tumor growth and metastasis formation.

CONCLUSION

This was an attempt to outline the present status of our knowledge concerning the 'cellular' contribution to blood clotting activation. Some normal cells (platelets, mononuclear phagocytes) or pathological cells (cancer cells) may trigger blood coagulation, either via the classical intrinsic and extrinsic pathways or via alternative mechanisms. Available evidence suggests that cellular procoagulants may be involved in the activation of blood coagulation in vivo. Platelet coagulant activities appear of great importance in physiological hemostasis. These platelet properties and especially the clot-promoting activities of mononuclear phagocytes and of cancer cells appear to be involved also in certain types of thrombotic disease and/or DIC. One of the problems confronted by anyone attempting to understand hemostasis and thrombosis is the nature of trigger mechanisms in vivo. A better understanding of the mechanisms modulating cellular procoagulants in vivo will possibly contribute to solving this enigma.

REFERENCES

1. E.M. Bevers, P. Comfurius, J. van Rijn, H.C. Hemker, and R.F.A. Zwall, Generation of prothrombin-converting activity and the exposure of phosphatidylserine at the outer surface of platelets, Eur. J. Biochem., 122:429 (1982).
2. P.W. Majerus and J.P. Miletich, Relationship between platelet and coagulation factors in hemostasis, Ann. Rev. Med., 29:41 (1978).
3. J.P. Miletich, C. Jackson, and P.W. Majerus, Properties of the factor X_a binding site on human platelets, J. Biol. Chem., 235:6908 (1978).
4. B. Østerus, S.I. Rapaport, and K.K. Lavine, Factor V activity of platelets: evidence for an activated factor V molecule and for a platelet activator, Blood, 49:819 (1977).
5. J.P. Miletich, W.H. Kane, S.L. Hofmann, N. Stanford, and P.W. Majerus, Deficiency of factor X_a-factor V_a binding sites on the platelets of a patient with a bleeding disorder, Blood, 54:1015 (1979).
6. P.N. Walsh, Platelet coagulant activities and hemostasis: a hypothesis, Blood, 43:597 (1974).

7.   P.N. Walsh and J.H. Griffin, Contributions of human platelets
     to the proteolytic activation of blood coagulation factors
     XII and XI, Blood, 57:106 (1981).

8.   G.P. Tuszynski and P.N. Walsh, Platelet membrane factor XI
     substitutes for plasma factor XI, Thromb. Haemost., 42:36
     (1979).

9.   P.N. Walsh, D.C.B. Mills, and F.I. Pareti, Hereditary giant
     platelet syndrome. Absence of collagen-induced coagulant
     activity and deficiency of factor XI binding to platelets,
     Br. J. Haematol., 29:639 (1975).

10. P.N. Walsh, P.H. Rogers, and V.J. Marder, The relationship
     of platelet coagulant activities to venous thrombosis fol-
     lowing hip surgery, Br. J. Haematol., 32:421 (1976).

11. P.N. Walsh, F.I. Pareti, and J.J. Corbett, Platelet coagulant
     activities and serum lipids in transient cerebral ischemia,
     N. Engl. J. Med., 295:854 (1976).

12. P.N. Walsh, S. Murphy, and W.E. Barry, The role of platelets
     in the pathogenesis of thrombosis and hemorrhage in patients
     with thrombocytosis, Thromb. Haemost., 38:1085 (1977).

13. N. Semeraro and J. Vermylen, Evidence that washed human
     platelets possess factor X activator activity, Br. J.
     Haematol., 36:107 (1977).

14. E.M. van Wijk, L.N. Kahlè, and J.W. Ten Cate, Factor X ac-
     tivation by washed human platelets, Thromb. Haemost., 42:56
     (1979).

15. S. Cortelazzo, M. Colucci, T. Barbui, E. Dini, and N. Semeraro,
     Reduced platelet factor X-activating activity. A possible
     contribution to bleeding complications in polycythemia vera
     and essential thrombocythemia, Haemostasis, 10:37 (1981).

16. N. Semeraro, D. Locati, and M. Colucci, Leukocyte-dependent
     platelet activation: an alternative pathway for initiation
     of blood clotting in inflammation, Agents Actions, 11:553
     (1981).

17. R.L. Edwards and F.R. Rickles, The role of monocyte tissue
     factor in the immune response, in: "Lymphokine Reports,"
     Vol. I, E. Pick, ed., Academic Press, N.Y. (1980).

18. R.P.A. Rivers, W.E. Hathaway, and W.L. Weston, The endotoxin-
     induced coagulant activity of human monocytes, Br. J.
     Haematol., 30:311 (1975).

19. E. Hiller, J.G. Saal, and G. Riethmuller, Procoagulant activity
     of activated monocytes, Haemostasis, 6:347 (1977).

20. H. Prydz and A.C. Allison, Tissue thromboplastin activity of

human monocytes, Thromb. Haemost., 39:582 (1978).

21. G.A. Levy, B.S. Schwartz, and T.S. Edgington, The kinetic and metabolic requirements for direct lymphocyte induction of human procoagulant monokines by bacterial lipopolysaccharide, J. Immunol., 127:357 (1981).

22. G.A. Levy and T.S. Edgington, Lymphocyte cooperation is required for amplification of macrophage procoagulant activity, J. Exp. Med., 151:1232 (1980).

23. E. Amlye, T. Lyberg, A. Kaplun, Ø. Hetland, and H. Prydz, Thromboplastin activity of mouse peritoneal macrophages, Thromb. Res., 24:61 (1981).

24. C.L. Geczy and K.E. Hooper, A mechanism of migration inhibition is delayed-type hypersensitivity reactions. II. Lymphokines promote procoagulant activity of macrophages in vitro, J. Immunol., 126:1059 (1981).

25. R.V. Mayer and R.J. Ulevitch, The response of isolated rabbit hepatic macrophages (H-Mø) to lipopolysaccharide (LPD), Circulat. Shock, 8:165 (1981).

26. B.S. Schwartz and T.S. Edgington, Immune complex-induced human monocyte procoagulant activity, J. Exp. Med., 154:892 (1981).

27. R.L. Edwards and F.R. Rickles, The role of human T cells (and T cell products) for monocyte tissue factor generation, J. Immunol., 125:606 (1980).

28. T. Lyberg and H. Prydz, Lectin stimulation of tissue thromboplastin activity in human monocytes in vitro, Thromb. Haemost., 42:1574 (1979).

29. N. Semeraro, A. Biondi, R. Lorenzet, D. Locati, A. Mantovani, and M.B. Donati, Direct induction of tissue factor synthesis by endotoxin in human macrophages from diverse anatomical sites, Immunology, 50:in press (1983).

30. T. Lyberg, K. Nilsson, and H. Prydz, Synthesis of thromboplastin by U-937 cells, Br. J. Haematol., 51:631 (1982).

31. N. Semeraro, Interactions of platelets, leukocytes, and endothelium with bacterial endotoxin: possible relevance in kidney disorders, in: "Hemostasis, Prostaglandins and Renal Disease," G. Remuzzi, G. Mecca, and G. de Gaetano, eds., Raven Press, New York (1980).

32. N. Semeraro, M. Colucci, L. Mussoni, and M.B. Donati, Rat blood leukocytes, unlike rabbit leukocytes, do not generate procoagulant activity on exposure to endotoxin, Br. J. Exp. Path., 62:638 (1981).

33. D.M. Komp and M.H. Donaldson, Sepsis in leukemia and the Schwartzman reaction, Am. J. Dis. Child., 119:114 (1970).

34. S. Cortelazzo, P. Viero, T. Barbui, M. Colucci, and N. Semeraro, Reduced generation of procoagulant activity by endotoxin-stimulated mononuclear cells from patients with chronic myeloid leukaemia, Br. J. Haematol., 48:501 (1981).

35. S. Cortelazzo, P. Viero, T. Barbui, M. Colucci, and N. Semeraro, Impaired production of mononuclear cell procoagulant activity in chronic lymphocytic leukaemia, J. Clin. Path., 36:37 (1983).

36. R.G. Lerner, R. Goldstein, and J.C. Nelson, Production of thromboplastin (tissue factor) and thrombi by polymorpho-nuclear neutrophilic leukocytes adhering to vein walls, Thromb. Res., 11:11 (1977).

37. C.L. Miller, G. Graziano, R.C. Lim, and M. Chin, Generation of tissue factor by patient monocytes: correlation to thromboembolic complications, Thromb. Haemost., 46:489 (1981).

38. M.B. Donati, A. Poggi, and N. Semeraro, Coagulation and ma-lignancy, in: "Recent Advances in Blood Coagulation," L. Poller, ed., Churchill Livingstone, Edinburgh (1981).

39. R.L. Edwards, F.R. Rickles, and M. Cronlund, Abnormalities of blood coagulation in patients with cancer. Mononuclear cell tissue factor generation, J. Lab. Clin. Med., 98:917 (1981).

40. R. Lorenzet, G. Peri, D. Locati, P. Allavena, M. Colucci, N. Semeraro, A. Mantovani, and M.B. Donati, Generation of pro-coagulant activity by mononuclear phagocytes: a possible mechanism contributing to blood clotting activation within malignant tissue, Blood, 62:271 (1983).

41. B.S. Schwartz, G.A. Levy, L.K. Curtiss, D.S. Fair, and T.S. Edgington, Plasma lipoprotein induction and suppression of the generation of cellular procoagulant activity in vitro, J. Clin. Invest., 67:1650 (1981).

42. G.F. Pineo, M.C. Brain, A.S. Gallus, J. Hirsh, M.W.C. Hatton, and E. Regoeczi, Tumor mucus production and hypercoagu-lability, Ann. N.Y. Acad. Sci., 230:262 (1974).

43. S.G. Gordon, J.J. Franks, and B.J. Lewis, Comparison of pro-coagulant activity in extracts of normal and malignant human tissue, J. Natl. Cancer Inst., 62:773 (1979).

44. L. Curatolo, M. Colucci, A. Poggi, L. Morasca, M.B. Donati, and N. Semeraro, Evidence that cells from experimental tumors can activate coagulation factor X, Br. J. Cancer, 40:228 (1979).

45. M. Colucci, L. Curatolo, M.B. Donati, and N. Semeraro, Cancer
     cell procoagulant activity: evaluation by an amidolytic assay,
     Thromb. Res., 8:589 (1980).
46. S.G. Gordon and B.J. Lewis, Comparison of procoagulant ac-
     tivity in tissue culture medium from normal and transformed
     fibroblasts, Cancer Res., 38:2467 (1978).
47. S.G. Gordon and B.A. Cross, A factor X activating cysteine
     protease from malignant tissue, J. Clin. Invest., 67:1655
     (1981).
48. H.R. Gralnick and E. Abrell, Studies on the procoagulant and
     fibrinolytic activity of promyelocytes in acute promyelocytic
     leukaemia, Br. J. Haematol., 24:88 (1973).
49. M. Colucci, R. Giavazzi, G. Alessandri, N. Semeraro, A.
     Mantovani, and M.B. Donati, Procoagulant activity of sarcoma
     sublines with different metastatic potential, Blood, 57:733
     (1981).

ANTITHROMBIN III AND OTHER NATURAL INHIBITORS

OF COAGULATION

Tiziano Barbui and Guido Finazzi

Divisione di Ematologia
Ospedali Riuniti di Bergamo

INTRODUCTION

The presence of naturally occurring inhibitory proteins appears to be one of the major systems for controlling the activation sequence in coagulation and for limiting thrombin formation. Plasma proteins which have been characterized as inhibitors of coagulation enzymes and cofactors are listed in Table 1.

Direct evidence regarding the role of these inhibitors in vivo is provided by experiments in nature in which congenital deficiency of inhibitors is associated with human disease. A familial tendency to thrombosis has been reported in inherited deficiency of antithrombin (AT) and protein C, suggesting that these two proteins play a major role in controlling the blood coagulation mechanism. Alpha-2-macroglobulin, alpha-1-antitrypsin, and Cl inhibitor have a minor physiological relevance in hemostasis. Relatively little is known about "new" endogenous inhibitors, such as heparin cofactor II (1).

ANTITHROMBIN

The Roman numeral classification system of antithrombins distinguishes between AT I (thrombin-absorbing effect of fibrin), AT II (heparin cofactor), AT III (slow inactivation of thrombin), and other types without indicating a significant physiological role.

Table 1. Natural Inhibitors of Coagulation

Inhibitor	Plasmatic Concentration mg/100 ml	Molecular Weight	Disease Associated with Inherited Deficiency
Antithrombin	24±2	65,000	Familial thrombosis
Protein C	0.4	62,000	Familial thrombosis
Alpha-2-Macroglobulin	260±70	725,000	None
Alpha-1-Antitrypsin	290±45	54,000	Pulmonary emphysema Hepatic cirrhosis
Cl Inactivator	24±3	104,000	Hereditary angio-neurotic edema
Heparin Cofactor	9	65,000	Inherited deficiency not described

Since a single plasma protein exerts both antithrombin II and III activities and the other "antithrombins" are not enzyme inhibitors, this nomenclature is no longer necessary. In accordance with recent publications, the term antithrombin (AT) will be used in the present paper.

Biochemical Aspects

Structure AT is a single chain glycoprotein migrating in standard electrophoresis to alpha-2 position. The primary structure of the molecule has nearly been completely determined: it contains approximately 424 aminoacids, 3 disulphide bridges, and 4 gluco-samine-based oligosaccharide side chains.

Interaction with Serine Proteases AT inhibits, at different rates, all the serine proteases of the coagulation cascade forming a 1:1 stoichiometric complex enzymatically inactive. AT appears to act as a "pseudosubstrate" for proteases. Complex formation may be due to a strong interaction between the active site of serine of

the enzyme and a specific Arg-Ser bond in the carboxy-terminal region of the inhibitor. The AT cleaved by protease may either dis-
sociate as free modified AT or form a final tight inactive AT-pro-
tease complex (2).

   Heparin and Heparin Fractions   The interaction between AT and
clotting enzymes is catalyzed by the presence of heparin which accelerates the rate without altering the stoichiometry of the reaction. The molecular mechanism with which heparin exerts its catalytic role has not been fully established. It seems likely that
the most efficient inhibition of thrombin takes place when both
AT and thrombin bind to the same heparin molecule. Once the AT-
thrombin complex is formed, heparin is released for further binding.
However, the mechanism of inhibition of factor Xa appears to be
different. Since binding of factor Xa to heparin is much weaker than
that of thrombin, it is probable that, unlike thrombin, direct binding of factor Xa to heparin is relatively unimportant.

   Differences between thrombin and factor Xa inhibition are
further stressed by studies of heparin fractions. Heparin can be
fractionated, with respect to antithrombin affinity, in high affinity (HA) and low affinity (LA) fractions and, with respect to
molecular weight, in high and low molecular weight. A direct relationship between molecular weight, binding to antithrombin, thrombin-
inhibiting capacity, and overall anticoagulant activity has been
recently demonstrated (3). In addition, high molecular weight
heparin fractions are considerably more active in inducing platelet
aggregation. However, low molecular weight fractions are more efficient in factor Xa inactivation. Based on these findings, improvement in the safety of haparin therapy may be achieved by utilizing mucopolysaccharide fractions with low molecular weight and high
affinity to antithrombin. These preparations would be expected to
minimize anticoagulant potency, via factor Xa inactivation, with a
minimum of undesirable effects on overall clotting and platelet function.

   Endogenous Glycosaminoglycans   Although the presence of endogenous heparin in human blood has not been demonstrated convincingly,
closely allied glycosaminoglycans such as heparan sulfate and derma-
tan sulfate have been found on the surface of endothelial cells.
Furthermore, it has been shown that human endothelial cells in
culture synthesize AT. The extent to which the action of AT can be

Table II. Classification of Congenital AT Deficiency Variants

Variant	Biological Assay	Immunological Assay	DCIE with Heparin	Pathophysiological Basis
Ia	Reduced	Reduced	Normal	Reduced synthesis (6)
IIb	Reduced	Reduced	Abnormal	Reduced synthesis + reduced affinity to heparin (?) (7)
IIa (het-erogeneous)	Reduced	Normal	Normal	Abnormal binding to thrombin (?) (8,9,10,11)
IIb	Reduced	Normal	Abnormal	Reduced affinity to heparin (12)

accelerated by endogenous glycosaminoglycans is, at present, uncertain. However, some preparations of heparan sulfate bind to AT
and have a relatively potent antifactor Xa activity. Drugs which
release endogenous glycosaminoglycans from endothelial cells represent a recent attractive approach to the prophylaxis of deep vein
thrombosis (4).

CLINICAL ASPECTS

Assay System In principle, two groups of methods for the
quantitative determination of AT are available. Measurement of the
concentration of AT antigen and determination of inhibitory activity
towards thrombin and factor Xa. The presence of an abnormal AT
molecule can give misleading results with immunological assays.
Furthermore, a biological assay is preferred to an immunological
one in monitoring replacement therapy with AT concentrates. The
introduction of synthetic chromogenic substrates has greatly facilitated the assay procedures. At present, an amidolytic biological
assay is commonly used in most laboratories for routine estimation
of AT. According to Sas et al. (5) qualitative information about
AT molecule can be obtained with double-cross immunoelectrophoresis
performed both in presence and in absence of heparin and by purification of the molecule by affinity chromatography on heparin-sepharose.

Congenital AT Deficiency is associated with a severe tendency
to venous thrombosis. It is inherited as an autosomal dominant
trait and, in the families studies, plasma level of AT is approximately 50% normal. Affected individuals share certain general
clinical characteristics. The mean age is 30 $\pm$ 14 years and a
familial history of venous thrombosis is common. The sites of
thrombosis are usually the veins of the lower half of the body, often
with complicating pulmonary embolism. Additional exogenous factors
including surgery, pregnancy, trauma, and infections often "trigger"
the thrombotic event. A genetic heterogeneity of congenital AT
deficiency is well known. A classification of the reported variants,
with respect to biological and immunological values and AT mobility
in double-cross immunoelectrophoresis in the presence of heparin,
is shown in Table II.

We have recently described an Italian family with a strong
history of familial thrombosis (8,11). Immunological AT levels,
heparin cofactor values, and double-cross immunoelectrophoretic

Table III. AT Pattern in 9 Family Members

Patient	Age (Y)	AT Assay Biological (%)	AT Assay Immunol. (%)	DCIE with Heparin (Plasma)
Propositus	30	62	105	Normal
I-1	56	94	90	
I-4	53	64	110	
I-5	47	68	140	
II-1	22	88	90	
II-2	27	103	104	
II-3	18	60	110	Normal
III-1	5	102	105	
III-3	5	65	107	
Normal range		85-115	85-125	

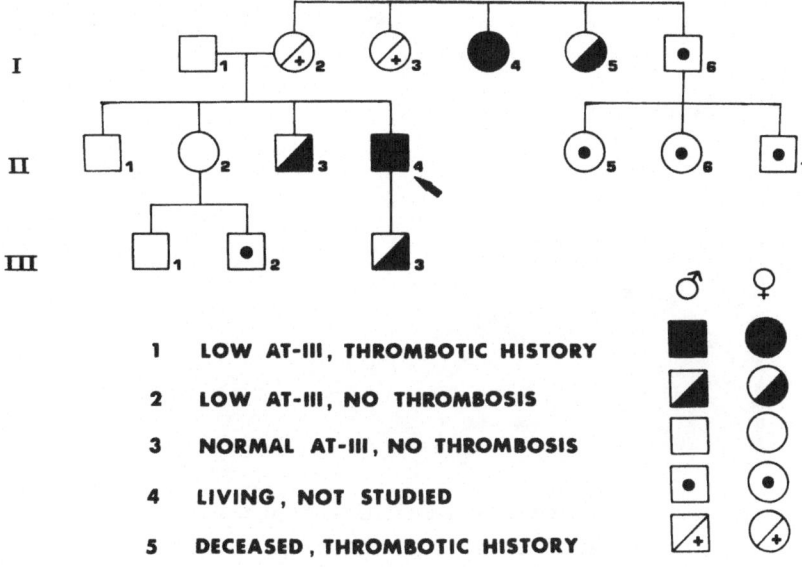

Fig. 1. Pedigree of the family.

mobility of AT in heparin/agarose were consistent with a type IIa
AT congenital deficiency (Fig. 1 and Table III).

This abnormal AT, called AT "Vicenza," has been more extensively
studied. Progressive thrombin inactivation (i.e. in absence of
heparin) and factor Xa inhibition, both in presence and in absence
of heparin, were all reduced to a similar degree (about 50% normal).
Interestingly, DCIE mobility of AT in presence of heparin, which
was normal in plasma, was clearly abnormal in serum. The fastest
more anodal peak, expression of residual active AT, was significantly
reduced compared to a normal serum. Similarly, slow peaks, expression of complexed or proteolyzed inactive AT, were proportionally
increased (Fig. 2). A similar abnormality was also found in plasma
artificially clotted with thrombin (Fig. 3).

Furthermore, AT "Vicenza" has been purified by means of affinity
chromatography on heparin-sepharose. Chromatographic pattern con-
clusively showed that heparin affinity of this dysfunctional AT was
normal (Fig. 4). Functional alteration of AT "Vicenza" is, most
likely, due to an abnormal binding to thrombin and factor Xa.

Acquired AT Deficiency has been described in a number of
diseases including severe liver disease with impaired hepatocyte

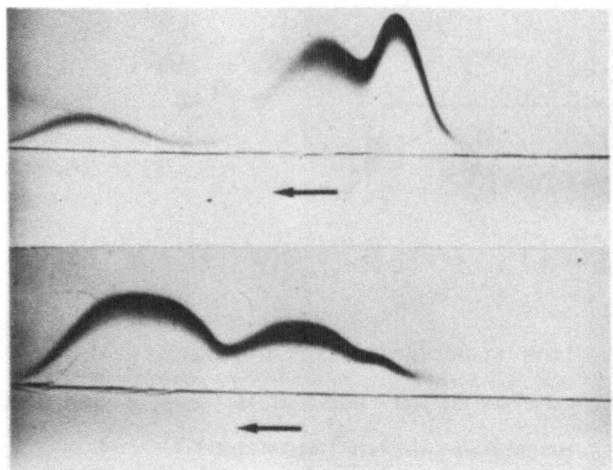

Fig. 2.   Double-cross immunoelectrophoresis
in heparin/agarose gels of normal
(bottom) and patient (top) serum.

function, nephrotic syndrome, disseminated intravascular coagulation, and acute venous thrombosis. A significant fall in plasma
level of AT has been also reported during heparin infusion and
during 1-asparaginase therapy. AT decrease during hormonal contra-
ceptive treatment appears to be related to the estrogen content
of the therapy. Since in most of these clinical situations procoagulant factors are also reduced, hemostatic balance is commonly
less impaired than in congenital AT deficiency and thrombotic risk
is less severe. A mathematical model concerning the procoagulant
factors/AT ratio has been recently proposed to predict the potential
thrombin generated in a plasma system in steady state and the potential thrombotic risk (13).

   AT Replacement could be a rational therapeutical approach to
the patient with congenital and/or acquired AT deficiency, with the
aim of treating or preventing venous thromboembolism and disseminated
intravascular coagulation. Recently, highly concentrated AT preparations have become available, but clinical experience is still limited. Efficient prevention of thrombotic episodes has been obtained
in congenital AT deficiency in critical clinical situations in which
oral anticoagulant therapy was inadequate or hazardous (14). Furthermore, good results have been reported in the normalization of
hemostatic function in patients with DIC and very low AT levels
(15). It is common opinion that clinical usefulness of AT concen-

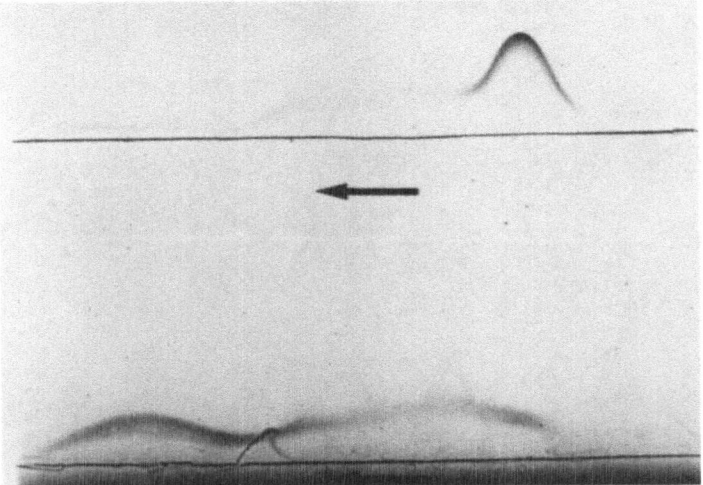

Fig. 3. Double-cross immunoelectrophoresis
in heparin/agarose gel of normal
(bottom) and patient (top) plasma
clotted with 120 U/ml of thrombin.

trates is probable, but that a larger number of patients must be
studied for a definitive evaluation of the possibilities of the
AT replacement therapy.

PROTEIN C

Protein C, a vitamin K-dependent plasma glycoprotein, is the
zymogen of a serine protease, activated protein C. Protein C may
be activated in vivo by trace amounts of thrombin and in vitro by
trypsin and by Russell's viper venom. Activation of protein C is
greatly accelerated by a cofactor from endothelial cells.

Activated protein C exhibits a potent anticoagulant activity
through a selective inhibition of factors Va and VIIIa by limited
proteolysis. Furthermore, activated protein C inhibits prothrombin-
forming activity of washed platelets and generates fibrinolytic
activity in blood by increasing levels of circulating plasminogen
activator. An inhibitor of activated protein C is present in normal
human plasma, and a deficiency of this inhibitor has been proposed
to be the basis of combined factor V/VIII deficiency disease (16).

The clinical relevance of protein C is strongly suggested by

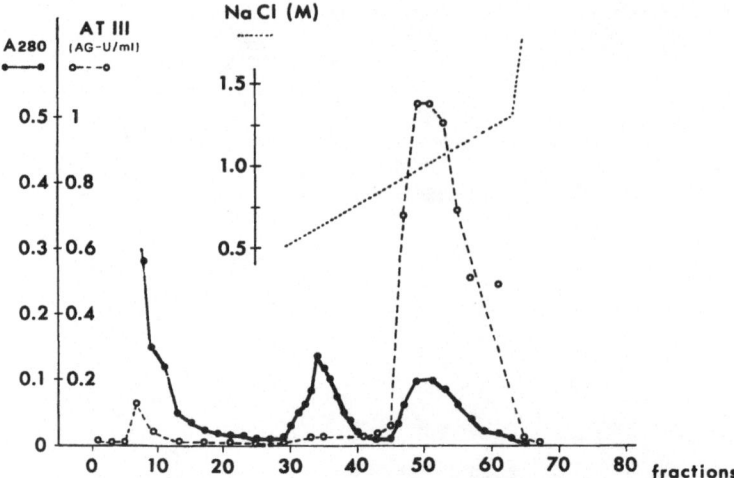

Fig. 4. Purification of AT "Vicenza" by
          affinity chromatography on heparin-
          sepharose.

reports that congenital deficiency of this inhibitor is associated
with a familial history of recurring venous thrombosis (17,18).
In addition, reduced levels of protein C have been described in
acquired clinical conditions associated with an increased tendency
to thrombosis, such as disseminated intravascular coagulation and
the period immediately following surgery (19).

     At present, only immunological systems are available for meas-
uring protein C concentration in plasma. A better understanding of
the importance of this inhibitor in congenital and acquired thrombophilic conditions is expected when biological assays systems will
be available.

REFERENCES

1.   D.M. Tellefsen, D.W. Majerus, and M.K. Blank, Heparin cofactor
        II. Purification and properties of a heparin-dependent
        inhibitor of thrombin in human plasma, J. Biol. Chem.,
        257:2162-2169 (1982).
2.   W.W. Fish and I. Bjork, Release of a two-chain form of anti-
        thrombin from the antithrombin-thrombin complex, Eur. J.
        Biochem., 101:31-38 (1979).
3.   S. Radoff and I. Danishefsky, High-affinity heparin: chain
        length, affinity for antithrombin and anticoagulant

activity, Arch. Biochem. & Biophy., 215:163-170 (1982).

4. D.P. Thomas, J.W. Barrowcliffe, R.E. Merton, J. Stocks, J. Dawes, and D.S. Pepper, In vivo release of antiXa clotting activity by a heparin analogue, Thromb. Res., 17:831-840 (1980).

5. G. Sas, D.S. Pepper, and J.D. Cash, Plasma and serum antithrombin III differentiation by crossed immunoelectrophoresis, Thromb. Res., 6:87-91 (1975).

6. D.R. Ambruso, B.D. Leonard, R.D. Bies, L, Jacobson, W.E. Hathaway, and E.B. Reeve, Antithrombin III deficiency: decreased synthesis of a biochemically normal molecule, Blood, 60:78-83 (1982).

7. G. Sas, I. Peto, D. Banhegyi, G. Blasko, and G. Domjan, Heterogeneity of the "classical" antithrombin III deficiency, Thromb. & Haemost., 43:133-136 (1980).

8. T. Barbui and F. Rodeghiero, Hereditary dysfunctional antithrombin III, Thromb. & Haemost., 45:97 (1981).

9. P.J. Sorensen, G. Sas, I. Peto, G. Blasko, T. Kremmer, and A. Samu, Distinction of two pathologic antithrombin III molecules: antithrombin III "Aalbor" and antithrombin III "Budapest," Thromb. Res., 26:211-219 (1982).

10. C. Boyer, A. Tripodi, M. Wolf, P.M. Mannucci, and M.J. Larrieu, Antitrombina III Milano: una nuova variante qualitativa. Atti del VII Congresso Nazionale della Società Italiana per lo studio dell'Emostasi e delle Trombosi, Roma 18-20 Ottobre, p. 58 (1982).

11. T. Barbui, G. Finazzi, F. Rodeghiero, and E. Dini, Immunoelectrophoretic evidence of a thrombin-induced abnormality in a new variant of hereditary dysfunctional antithrombin III (AT III 'Vicenza'), Br. J. Haematol., (in press).

12. T.H. Tran, H. Bounameaux, C. Bondeli, H. Honkanen, G.A. Marbert, and F. Duckert, Purification and partial characterization of a hereditary abnormal antithrombin III fraction of a patient with recurrent thrombophlebitis, Thromb. & Haemost., 44:87-91 (1980).

13. E.B. Reeve, Steady state relations between factors X, Xa, II, IIa antithrombin III and alpha-2-macroglobulin in thrombosis, Thromb. Res., 18:19-31 (1980).

14. P.M. Mannucci, C. Boyer, M. Wolf, A. Tripodi, and M.J. Larrieu, Treatment of congenital antithrombin III deficiency with concentrates, Br. J. Haematol., 50:531-535 (1982).

15. H.G. Schipper, C.S.P. Jenkins, L.H. Kahle, and J.W. Ten Cate,

Antithrombin III transfusion in disseminated intravascular coagulation, Lancet, ii:854–856 (1978).

16. R.A. Marlar and J.H. Griffin, Deficiency of protein C inhibitor in combined factor V/VIII disease, J. Clin. Invest., 66: 1186–1189 (1980).

17. J.H. Griffin, B. Evatt, T.S. Zimmermann, A.J. Kleiss, and C. Wideman, Deficiency of protein C in congenital thrombotic disease, J. Clin. Invest., 68:1370–1373 (1981).

18. R.M Bertina, A.W. Broekmans, I.K. Van Der Linden, and K. Mertens, Protein C deficiency in a Dutch family with thrombotic disease, Thromb. & Haemost., 48:1–5 (1982).

19. P.M. Mannucci and S. Viganò, Deficiencies of protein C, and inhibitor of blood coagulation, Lancet, ii:463–467 (1982).

ON THE EVALUATION OF THE THROMBOTIC RISK

G. Mariani, G.D. Di Nucci, A. Chistolini,
M. Motta, and F. Mandelli

Institute of Hematology
University of Rome
Via Chieti 7
Rome, Italy

The evaluation of the thrombotic risk can be done using a probabilistic approach. This is due to a new application of epidemiology: the field of perspective studies. The tools of this new application of the old discipline deal with the study of several parameters in a given population, and subsequently the application to individual patients of the statistical data obtained from the groups. In this context, the thrombotic risk factor is nothing more than a statistical predictor, that is, an attribute that occurs more frequently in persons with disease than in control people. The application of the epidemiological studies can be very useful, as only in a small number of situations the relationship between clinical and laboratory events on one hand and thrombosis on the other is fairly constant.

The problem is a very important one as thromboembolism represents the first cause of death in industrialized countries. Initially, the phenomenon may appear fairly benign and the lethal event can manifest itself only after repeated thrombotic complications or following its sequels, but whatever the thrombotic event, it always causes disability which tends to progress unless an effective treatment is carried out.

Another relevant aspect deals with the anti-thrombotic drugs:

heparin, oral anticoagulants, and thrombolytic drugs. All of these
drugs had been used well before the introduction of the so-called
tests for thrombosis had reached an acceptable diagnostic credibility.
These drugs, however, may carry an important hemorrhagic risk.
Because of this, the general consensus has been to increase efforts
to acquire a deeper knowledge of the event to be fought before
giving drugs that are potentially dangerous. In this sense, the
approaches have been essentially two: 1) the epidemiological-
anamnestic approach and 2) the laboratory approach.

EPIDEMIOLOGICAL APPROACH

    The risk factors that have been derived from the epidemiological
studies may be divided into two categories: a) conventional (or
generic) risk factors that acquire importance mainly if associated
to specific risk factors or to other risk factors of the same
category, and b) specific risk factors.

    The most known risk factors that can be categorized as
conventional are: age, sex, smoking habit, blood group, sedentary
life, obesity, incorrect dietary habits, increased blood pressure,
stress, hyperuricemia, blood sugar, and positive family history.

    It is known that in advanced age the hemostatic balance is
overweight in the sense of the hypercoagulability; it has been
reported that fibrinogen and factors V, VII, VIII, XI, X, and XI
do increase with age, some of them in both sexes, some in only one
(1). Antithrombin III (AT III) diminishes in elderly men and in
elderly women (2). There are records demonstrating that increased
incidence of thrombosis in smokers in both medical and surgical
wards. Blood groups other than O can also be considered a risk
factor, due to the fact that people bearing these blood groups
show increased levels of factor VIII (3).

    Cigarette smoking has been frequently found to be associated
with various thrombotic events (coronary, venous, arterial), but
the strength of association greatly depends on the presence of
other risk factors. Smoking has been proved to increased platelet
adhesiveness as well as aggregability (4), but this is true mainly
in young people.

    Increased blood pressure has frequently been found in association with cardiovascular or arterial thrombotic events. It is still a matter of some debate wheter the systolic or diastolic pressures are of prevalent importance. Increased blood pressure, however, can be considered a risk factor for thrombosis only if related to other risk factors, such as diabetes, obesity, sedentary life, etc. (5).

Stress may also cause an increase of factor VIII, together with an increase of fibrinolytic activity (this applies in the opposite sense). Obesity and sedentary life are both risk factors; they can be related to wrong dietary habits and to lipoprotein abnormalities. In fact, a high intake of saturated fatty acids increases platelet aggregation (6) and induces a reduction of AT III levels (7). Lipids such as cholesterol are implicated in atherogenesis (3). In hypertension, high levels of factor VIII have been described, together with a shortening of platelet half-life.

Increased factor VIII antigen (8) and fibrinogen levels as well as a reduction of AT III (9), increased platelet aggregation (10), and depression of fibrinolytic activity (11) have all been reported even more frequently in diabetes.

Pregnancy can also be considered a conventional risk factor; during pregnancy (mainly in the 3rd trimester) increased levels of factors VII, VIII, X, and fibrinogen (1) have been observed, but the thrombotic complications do not exceed 0.5% before and 0.1% after parturition; in most of the cases, risk factors other than pregnancy are associated (12).

The risk factors that might be identified as "specific" deal with those clinical situations during the course of which an increased incidence of thrombotic episodes have been demonstrated, even independently of the presence of other conventional risk factors. The most important risk factors "specific" for thromosis are listed in Table 1.

The thromboembolic complications in myocardial infarction are partly due to the coronary thrombosis itself which may be related to blood coagulation changes, in part due to the presence of other risk factors, such as stress, obesity, smoking habit, and immobilization. The incidence of thrombosis, assessed with the ^{125}I-labeled fibrinogen, in acute myocardial infarction (AMI) varies

Table 1. Specific Risk Factors for Thrombosis

Myocardial infarction
Surgery (hip, abdominal)
Prostetic heart valve
Chronic heart failure
Malignancies (lung, pancreas, prostate, tumors
    secreting mucin)
Myeloproliferative disorders (polycythemia vera,
    thrombocythemia, sickle cell anemia,
    nocturnal paroxysmal hemoglobinuria)
Burns and trauma
Infections (Gram-)
Previous thrombotic events

from 20 to 50% (13,14,15). Thrombosis occurs within the first
48-72 hours (15). In myocardial infarction, increased levels of
FpA have been found. Fibrinogen as well as factor VIII are frequently increased, but in this case it is difficult to ascertain whether
this is due to the "acute phase" or to the coronary thrombosis
(provided these two aspects can be kept separate). The presence of
artificial valves indeed represents an important risk factor for
thromboembolic complications; in fact, an incidence of 10% thrombotic
events has been reported during the first year of anticoagulant
therapy (16).

In surgery, the thrombotic risk can be partly due to local
factors (manipulation of veins during the surgical procedures), or
to general changes due to anesthesia and immobilization. After the
first postoperative day, and until the end of the first postoperative
week, several well defined changes of blood coagulation occur, all
of them concurring to create a hypercoagulable state: reduced
fibrinolytic activity, low plasminogen and AT III levels, protein
C reduction, and increase of factor VIII and fibrinogen levels (1).
This explains the high incidence of deep vein thrombosis (DVT) and
pulmonary embolism (PE) during the week following the surgical
operation. The thromboembolic complications are rather frequent
in patients affected by malignancies. Apart for the possibility of
specific activation of clotting that can be promoted by some tumor
cells (promyelocytes, prostatic cancer cells, and cell secreting
mucin), in patients with malignancies, an increase of fibrinogen
(17), factor VIII (17,18a), and platelet count is frequently found.

These changes are accompanied by a reduction of AT III (18) and of plasminogen (19).

Some infections are frequently associated with thromboembolic complications. This is true mainly when the infections are sustained by Gram-negative germs. In this context, endotoxin plays a very important role in cooperation with leukocytes (20), platelets, and plasma factors of the intrinsic pathway. In myeloproliferative disease, thrombosis can be a frequent complication. In thrombocythemia, thrombosis occurs more frequently than bleeding (21). In these diseases, the high platelet count and the blood hyperviscosity play a very important role; however, plasma procoagulant factors are also involved which give further explanation to the tendency to thrombosis (22).

Deep vein thrombosis and pulmonary embolism have been found to occur with high frequency in burns (29%) (24) and in severely traumatized patients (up to 50%). The frequency increases if survival goes beyond two weeks (23).

In sickle cell disease, thrombotic events are frequently seen. In patients not in crisis, hemostatic abnormalities can be found, characterized by elevated fibrinogen, factor VIII, and platelet count (25).

Some drugs have also demonstrated an increase in the risk of thrombosis with different mechanisms: 6-azauridine, acetrizoate, ristocetin, estrogens, the estrogen-progestative association (the pill) (1).

Lastly, we shall consider those situations in which the relationship with thrombosis is very close. AT III congenital deficiency represents one of the few pathological conditions in which the risk of thrombosis can be clearly defined on biochemical ground. AT III congenital deficiency is almost as frequent as hemophilia (26). At the age of 50, nearly all the AT III deficient patients have experienced at least one thrombotic event (27), mainly, if not exclusively, on the venous side; although in about 1/3 of the patients the thrombotic event occurs spontaneously, a precipitating condition, such as delivery, pregnancy, surgery, trauma, or septicemia, facilitates the occurrence of thrombosis (27). From the study of the first families, a tendency of thrombosis seems

to be equally strong in protein C congenital deficiency. Yet,
another clinical situation in which the thrombotic event is almost
constant is the DIC which occurs during the course of acute
promyelocytic leukemia (APL). In this case, the stimulus to
thrombosis is so strong that a thrombo-hemorrhagic syndrome results.

In conclusion, as far as the epidemiological risk factors are
concerned and apart from those diseases that are directly related
to the thrombotic event, none of the mentioned thrombotic risk
factors may be related to a certain tendency to thrombosis either
in normal subjects or in pathological conditions. To each of these
risk factors an intrinsic weight should be attributed. This would
be important in quantifying the thrombotic risk a priori. Considering the lack of a certain thrombotic risk in these clinical
situations, the evaluation of the risk may be improved by adding
suitable laboratory findings. The help that the laboratory can give
in the evaluation of the thrombotic risk deals with the three
following aspects: firstly, evaluation of pro-coagulant proteins,
secondly, evaluation of physiological inhibitions, and thirdly,
assessment of the fibrinolytic activity.

Procoagulant proteins are those proteins, serine proteases or
non-serine proteases, of platelet origin or plasma proteins that
may imbalance the hemostatic mechanism towards thrombosis. As the
changes dealing with these proteins may be different in different
clinical situations, we will refer to the following: a) deep vein
thrombosis and pulmonary embolism, b) arterial thrombosis, c) disseminated intravascular coagulation.

Deep vein thrombosis and pulmonary embolism may occur even in
patients who are congenitally deficient of factors VII and XII.
This is important because these two factors are presently considered
the main initiators of blood clotting. Thromboembolic complications
may also occur when functional fibrinogen changes are present, such
as dysfibrinogenemias. In the pathogenesis of venous thrombosis,
the hemodynamic components seem to prevail mainly in the first
stages of the process and so as a consequence. If we need to have
markers for vein thrombosis from blood samples, we have to utilize
those tests which are believed to be contemporary or which follow
the thrombosis itself, such as the release of fibrinopeptide A
(FpA), which gives indirect evidence of the presence of thrombin.
FpA has a very short $t_\frac{1}{2}$: 3-5 minutes. This implies that FpA

increased levels have to be interpreted as a consequence of an ongoing process. In fact, in 90% of the phlebographically demonstrated DVT, FpA levels are increased (28). Considering the difficulties that sometimes may be encountered in drawing blood to perform the FpA assay and the very short half-life of this peptide, it might be of interest to know the daily production of this fibrinogen derivative. This can be done by performing the assay of a 24 hour sample of urine (29). However, the most reliable test to perform in early diagnosis of venous thrombosis is still the ^{125}I-fibrinogen deposition test. When DVT is complicated by PE, increased levels of FDP and of the fibrinogen fragment E are frequently found as a consequence of an increased plasminic activity.

Arterial thrombosis occurs in an idraulic system characterized by a high speed flow. There are no solid data dealing with the initial mechanisms leading to arterial thrombosis. Some blood coagulation changes may be of importance. In ischemic heart disease (IHD), for instance, the presence of α-VII$_a$ in the blood stream may be considered a risk factor (3). Increased factor VIII as well as fibrinogen levels are found in IHD. An ongoing arterial thrombotic event may be diagnosed early in the blood stream if FpA levels are assayed together with the release of β-TG and PF$_4$.

β-TG as well as PF$_4$ were found to be increased in several arterial thrombotic disease (30,31,32). Malondialdehyde formation by platelets and thromboxane B$_2$ as markers of platelet lipids peroxidation have been assayed in several arterial thrombotic events (31,33). The whole platelet fate, assayed with several methods, such as platelet survival, in vivo platelet aggregates, and platelet aggregation have been studied in arterial thrombosis and in related disorders with interesting, but often non-consistant results.

In atherosclerosis, the importance of platelets in the pathogenesis of the arterial wall lesions has been stressed; however, recent data added the evidence that procoagulant proteins may also be important, as noticeable amounts of fibrinogen can also be found in the plaque. As regards to the risk factor and early diagnosis, the same observations put forward for arterial thrombosis may apply.

Disseminated intravascular coagulation in its acute clinical form represents the extreme limit to which a thrombotic stimulus

may imbalance hemostasis. Since DIC is a syndrome, several mechanisms due to a much wider number of noxae are thought to take place.
All these mechanisms cause or may lead to changes in blood coagulation that are characterized by fibrin (and platelet) deposition
in the microvasculature, which is followed by an activation of the
fibrinolytic system. The tests that may be used for an early
diagnosis of DIC are based on: 1) the study of the thrombin-
fibrinogen reaction, and 2) the study of the plasminic digestion
of fibrin. The assay of the platelet-derived proteins may be
considered subsidiary. The presence of thrombin may also be
revealed by increased FpA levels, whereas, the plasminic fibrin
digestion may be demonstrated by the assay of the $B\beta1-42$. This
peptide represents the early product of fibrin digestion (34).
In addition FDPs may be used to study the phenomenon. These tests
also have a therapeutical relevance as they may indicate that
heparin treatment should be given only when a prevalent thrombin
action is demonstrated by increased levels of FpA.

Alpha-globulins are physiological inhibitors which inhibit
circulating serine proteases or other procoagulant proteins.
Physiological inhibitors display a wide action as most of them
can inhibit more than one protease or factor. These inhibitors
are: AT III, α_2-macroglobulin, α_1-antitrypsin, Cl-inactivator,
and finally, protein C. AT III is the inhibitor that is responsible
for most of the antithrombotic action of the blood, and hence, is
the one which has been studied extensively. This is because its
congenital deficiency is correlated to a high incidence of thromboembolic complications. Low levels of AT III (below 70 U/dl) in
either acquired or congenital situations must be considered per se
a risk factor of thrombosis, which will acquire even more importance
if associated to other risk factors. The role of the other
physiological inhibitors in this context is less well known. In
addition, α_1-antitrypsin is a protein which rises during the acute
phase and, therefore, cannot be considered a valid marker for
thrombosis. As far as protein C is concerned, its peculiar role
of inhibitor, acting against procoagulant factors that are not
serine proteases (factors V and VIII), and the fact that in protein C
congenital deficiency a tendency to thrombosis has recently been
demonstrated, will certainly give us important data on the
physiological role of this vitamin K-dependent inhibitor.

The fibrinolytic system plays a very important role because

it induces lysis of the most important constituent of thrombi: fibrin. Reduced fibrinolytic activity predisposes undoubtedly to thrombosis, as demonstrated by the frequent occurrence of thrombotic complications in patients treated with high doses of antifibrinolytic agents, and by the changes the fibrinolytic system undergoes during the postoperative period (35). Reduced fibrinolytic activity was found in malignancies, pregnancy, peripheral arterial disease, diabetes, and obesity. It is possible to measure the whole fibrinolytic activity as well as the most important constituents of this system, namely, the extrinsic plasminogen activator, plasminogen itself, and α_2-antiplasmin.

We have referred only to some aspects that deal with the vast and important problems concerning the thrombotic risk. As thrombosis is certainly a multifactorial event, and as probably the mechanisms triggered off in different clinical occasions may be different, we cannot forecast the proposition of a single "test for thrombosis." In addition, it is difficult to separate the blood coagulation changes which are directly responsible for the thrombotic event from those that are contemporary to the thrombotic event, and finally those that represent a late effect from it. In consequence, it is difficult to give a test a predictive value, and hence, to label it as a risk factor. There are also kinetic and anatomical considerations which allow us to understand why it is so difficult to find a test that can be utilized as a marker for thrombosis. These considerations are: 1) serine proteases may be removed very rapidly from the circulation by physiological inhibitors, 2) their assay in peripheral blood may be minimized by the dilution effect, and 3) the reticuloendothelial system clears the plasma from these activated factors.

In conclusion, it is reasonable to forecast that in the near future the laboratory evaluation of the thrombotic risk will be performed on the basis of: a) very sensitive assay of plasma factors or peptides, b) tests based on dynamic evaluation of procoagulant proteins or physiological inhibitors, and finally c) tests reflecting platelet metabolic activities.

The last and most important problem deals with the combination of the risk factors for thrombosis. How much weight should be given to the epidemiological odds? How much weight should be given to the laboratory odds? The lack of definite knowledge of the

of the thrombotic events clearly account for the difficulties
dealing with the state of the art of the thrombotic risk.

REFERENCES

1. C.A. Owen and E.J.W. Bowie, Predisposing factors in thrombosis,
    in: "Thrombosis," H.C. Kwaan and E.J.W. Bowie, eds.,
    Saunders, Philadelphia (1982).
2. O.R. Ødegard, M.K. Fagerhold, and M. Lie, Heparin cofactor
    activity and antithrombin III concentration in plasma
    related to age, Scand. J. Haematol., 17:258 (1976).
3. T.W. Meade, Epidemiology of atheroma and thrombosis, in:
    "Haemostasis and Thrombosis," A.L. Bloom and D.P. Thomas,
    eds., Churchill Livingstone, Edingburgh (1981).
4. P.H. Levine, An acute effect of cigarette smoking on platelet
    function: a possible link between smoking and arterial
    thrombosis, Circulation, 48:619 (1973).
5. S.W. Rabkin and D.L. Sackett, Epidemiology of arterial
    thromboembolism, in: "Hemostasis and Thrombosis," R.W.
    Colman, J. Hirsh, V.J. Marder, and E.D. Salzman, eds.,
    Lippincott, Philadelphia (1982).
6. S. Renard, R. Morazain, L. McGregor, and F. Baudier, Dietary
    fats and platelet functions in relation to atherosclerosis
    and coronary heart disease, Haemostasis, 8:234 (1979).
7. M.L. Zucker, E.D. Gomperts, and D. Russel, Antithrombin
    functional activity after saturated and unsaturated fatty
    meals and fasting in normal subjects and some disease
    states, Thromb. Res., 15:37 (1979).
8. E.G. Lufkin, D.N. Fass, and W.M. O'Fallon, Increased von
    Willebrand factor in diabetes mellitus, Metabolism, 28:63
    (1979).
9. R.N. Banerjee, A.L. Sabri, V. Kumar, and M. Arya, Antithrombin
    III deficiency in maturity, onset diabetes mellitus and
    atherosclerosis, Thromb. Diath. Haemorrh., 31:339 (1974).
10. J.A. Colwell, R.M.G. Nair, P.V. Halushka, C. Rogers, A.
    Whetsell, and J. Sagel, Platelet adhesion and aggregation
    in diabetes mellitus, Metabolism, 28(Suppl. 1):394 (1979).
11. L.O. Almér and I.M. Nilsson, On fibrinolysis in diabetes mel-
    litus, Acta Med. Scand., 198:101 (1975).
12. E. McDevitt and B. Smith, Thrombophlebitis during pregnancy
    and the puerperium, in: "Thrombosis," S. Sherry, K.M.
    Brinkhous, and E. Genton, eds., Nat. Acad. Sci.,
    Washington (1969).

13. T.S. Murray, F.C. Fox, A.R. Lorimer, and T.D.V. Lowrie, Leg vein thrombosis following myocardial infarction, Lancet, ii:792 (1970).

14. B. Modan, E. Sharon, and N. Jelin, Factors contributing to the incorrect diagnosis of pulmonary embolic disease, Chest, 62:388 (1972).

15. B.J. Maurer, R.W. Ray, and J. Shillingford, Frequency of venous thrombosis after myocardial infarction, Lancet, ii:1385 (1970).

16. D.A. Barnhost, H.A. Oxman, D.C. Connolly, J.R. Pluth, G.K. Danielson, R.B. Wallace, and D.C. McGoon, Long-term follow-up of isolated replacement of the aortic and ventricle valve with the Starr-Edwards prothesis, Am. J. Cardiol., 35:10 (1975).

17. N.C.J. Sun, W.M. McAfee, G.J. Hum, and J.M. Weiner, Hemostatic abnormalities in malignancy, a prospective study of one hundred and eight patients. Part I - Coagulation studies, Am. J. Clin. Pathol., 71:10 (1979).

18. R. Losito, P. Beaudry, J.C. Valderrama, L. Cousineau, and B. Longpre, antithrombin III and factor VIII in patients with neoplasms, Am. J. Clin. Pathol., 68:258 (1977).

18a. S.J. Scialla, S.E. Speckart, M.S. Haut, and D.B. Kimball, Alterations in platelet surface sialyltransferase activity and platelet aggregation in a group of cancer patients with a high incidence of thrombosis, Cancer Res., 39:2031 (1979).

19. J.A.N. Rennie and D. Ogston, Fibrinolytic activity in malignant diasease, Am. J. Clin. Pathol., 28:872 (1975).

20. F.R. Rickles, J. Levin, J.A. Hardin, C.F. Barr, and M.E. Conrad, Tissue factor generation by human mononuclear cells: effect of endotoxin and dissociation of tissue factor generation from mitogenic response, J. Lab. Clin. Med., 89:792 (1977).

21. S. Hussain, J.M.S. Schwartz, S.A. Friedman, and S.A. Chua, Arterial thrombosis in essential thrombocythemia, Am. Heart J., 96:31 (1978).

22. A. Carvalho and L. Ellman, Activation of the coagulation system in polycythemia vera, Blood, 47:669 (1976).

23. S. Sevitt, Venous thrombosis in injured patients (with some observations on pathogenesis), in: "Thrombosis," S. Sherry, K.M. Brinkhous, E. Genton, eds., Nat. Acad. Sci., Washington (1969).

24. J.B. Coleman and F.C. Chang, Pulmonary embolism: an unrecognized event in severely burned patients, Am. J. Surg., 130:697 (1975).

25. N.E. Stathakis, A.G. Papayannis, and H. Papayotas, Hyper-
     coagulability and hypofibrinolysis in sickle cell disease,
     Blut, 31:355 (1975).

26. O.R. Ødegard and U. Abilgaard, Antithrombin III: critical
     review of assay methods. Significance of variation in
     health and disease, Haemostasis, 7:127 (1978).

27. E. Thaler and K. Lechner, Antithrombin III deficiency and
     thromboembolism, in: "Clinics in Haematology," C.R.M.
     Prentice, ed., W.B. Saunders Co. Ltd., London (1981).

28. I. Yudelman and J. Greenberg, Factors affecting fibrinopeptide
     A levels in patients with venous thromboembolism during
     anticoagulant therapy, Blood, 4:787 (1982).

29. N. Alkjaersig and A.P. Fletcher, Catabolism and excretion of
     fibrinopeptide A, Blood, 60:148 (1982).

30. B. Ruckinski, S. Niewiarowsky, P. James, D.A. Walz, and A.
     Budzinski, Antiheparin proteins secreted by human
     platelets. Purification, characterization, and radio-
     immunoassay, Blood, 53:47 (1979).

31. J.R. O'Brien, M.D. Etherington, R.D. Shuttleworth, and W.H.
     Calwell, Platelet function in acute myocardial infarction
     patients compared with controls, Thromb. Haemost., 44:
     96 (1980).

32. G. Cella, J. Zahavi, H.A. deHaas, and V.V. Kakkar, Beta-
     thromboglobulin, platelet production time, and platelet
     function in vascular disease, Br. J. Haematol., 43:127
     (1979).

33. R. Bizios, L.K. Wong, R. Vaillancourt, R.S. Lees, and A.
     Carvalho, Platelet prostaglandin endoperoxides formation
     in hyperlipidemias, Thromb. Haemost., 38:228 (1977).

34. H.L. Nossel, Coagulant proteins in thrombosis, in: "Hemostasis
     and Thrombosis," R.W. Colman, J. Hirsh, V.J. Marder, and
     E.W. Salzman, eds., Lippincott, Philadelphia (1982).

35. A.O. Mansfield, Alteration in fibrinolysis associated with
     surgery and venous thrombosis, Br. J. Surg., 59:754 (1972).

THE CORRELATION BETWEEN FATTY ACIDS AND

PLATELET FUNCTION

E. Cacciola, R. Musso, R. Giustolisi,
N.M. Catania, E. Cacciola, and R.R. Cacciola

University of Catania
Chair of Hematology
Ospedale Ferrarotto
Via S. Citelli
Catania, Italy

INTRODUCTION

It is well known that fatty acids (FA) show in vitro the capability to induce platelet aggregation when employed at optimal concentrations (1-4). On the other hand, there is evidence that among FA the arachidonic acid (AA) plays a key role in platelet metabolic pathway leading to the synthesis of strongly aggregating agents, such as prostaglandins (PG) and thromboxanes (TXA) (5-9). Moreover, it has been documented that some inducers can activate platelet membrane lipases (10-11), so that their aggregating effect would also be mediated by a lipometabolic activation (11-12). Such data, therefore, would demonstrate that a correlation between platelet membrane phospholipids (13-14) and aggregative capability degree of inducers could exist.

MATERIALS AND METHODS

Preparation of FA Solutions

FA (Nu Check Prep. Inc., Elysian, Minn., USA) were used as Na salts, 99% pure, and prepared as follows: a) saturated FA (s-FA): propionic acid (3:0), butyric acid (4:0), valeric acid (5:0), caprylic acid (8:0), capric acid (10:0), undecenoic acid (11:0), myristic acid (14:0), palmitic acid (16:0), stearic acid (18:0), nonodecilic acid (19:0), arachic acid (20:0), behenic acid (22:0), diluted in TRIS-buffer-saline solution (0/1 M, pH 8.5 (15); b) unsaturated FA (u-FA): erucic acid (22:1), arachidonic acid (20:4) in solution of 0.1 M Na_2CO_3 under nitrogen, pH 8.5, to prevent the oxidation (16), were stored at -70°C in aliquots and then dissolved at 37°C just before use. The FA solutions were prepared at 5×10^{-3} M concentration (stock solution), and employed at 0.4 mM final subcritical concentration (not aggregating) into 1 ml PRP sample.

Blood Collection and Preparation of Plasma

Venous blood was collected from healthy volunteers, aged 18-35 years, who had been fasting for 10 hours and had not taken any drugs during the previous week. Nine vol. of blood were collected into 1 vol. of trisodium citrate 0.11 M in plastic tubes. PRP and platelet-poor plasma (PPP) were immediately prepared by centrifugation at room temperature at 220 g. for 15 minutes and 2,400 g. for 30 minutes, respectively. PRP was adjusted to $2.5 \times 10^5/\mu l$ platelets by use of homologous PPP. The plasma samples were kept at 20°C and used within 4 hours after venipuncture. Platelet counts were performed using a phase contrast microscope.

Platelet Aggregation

Platelet aggregation was recorded at 37°C and 1.10^3 RPM according to the turbidimetric method of Born (17) in a dual channel aggregation 840 module (Elvi Logos, Milan, Italy) linked to Omniscribe recorder (Logos 176, Milan, Italy) by using 500 μl of PRP. The extent of aggregation was measured as percent light transmission, using the light transmission of PRP as 0% and that of PPP as 100%.

Aggregating Substances

Adenosine-5 diphosphate, sodium salt (ADP, Sigma Chem., St. Louis, Miss., USA) was dissolved in saline, stored at -70°C in aliquots at 10^{-3} (stock solution) and employed at 0.6×10^{-6} M critical concentration into PRP sample. L (-) epinephrine (Parke-

Davis) in 1% saline was employed at 1x10   M critical concentration
into PRP sample. Collagen (Horm Chemicals, Munich, West Germany)
prepared in acid buffer (SKF Horm 10,500), pH 2.5, was employed
at 1 μg final concentration in 1 ml PRP. Thrombin (Parke-Davis)
was dissolved in saline at 25 U/ml concentration (stock solution)
and stored at -70°C in small aliquots. 0.3 U final concentration
of thrombin was added to 1 ml PRP.

Test Procedure

All platelet aggregation experiments were carried out in
duplicate. The incubation mixture consisted of 500 μl PRP and
0.4 mM individual FA (not aggregating concentration) for 3 minutes,
under continuous stirring, at 37°C. Then platelet aggregation was
induced with the above mentioned inducers, respectively. After 5
minutes aggregation, the percentage of platelet aggregability was
recorded.

RESULTS

In control procedures individual FA, employed at subcritical
concentration, did not induce any aggregation, whereas, ADP,
epinephrine, collagen, and thrombin, when used alone at critical
concentration, determined minimal platelet aggregation. The
platelet aggregative curves were reversible in presence of ADP and
thrombin, and not reversible (with a secondary wave) in presence of
epinephrine. Normal lag period and normal aggregatory response
were observed in collagen-induced aggregation curves.

In PRP mixtures with individual FA (at subcritical concentration) and the above mentioned inducers (at critical concentration),
we observed a potentiating effect of platelet aggregation only in
presence of long carbon chains FA, both saturated and unsaturated,
unlike the short carbon chains FA. Such potentiating effect was
characterized by the appearance of a secondary wave in ADP and
thrombin normal curves. In epinephrine-induced platelet aggregation,
a disappearance of two phases with an enhanced aggregation rate
was observed, while a shortened lag period and increased aggregation
rate were noted in collagen-induced platelet aggregability
(Table I,II - Fig. 1).

Table I.   Effect of Subcritical Concentrations (0.4 mM) of FA in ADP-Induced
           (0.6 μM) Platelet Aggregation

Subcritical concentration (0.4mM) of FA with PRP on ADP-induced (0.6 μM) platelet aggregation	ADP-induced aggregation %	Behavior of aggregatory platelet response
(C 3:0)*	18 (20)§	Reversible
(C 4:0)	20 (20)	"
(C 5:0)	18 (18)	"
(C 8:0)	20 (18)	"
(C 10:0)	20 (20)	"
(C 11:0)	20 (20)	"
(C 14:0)	22 (18)	"
(C 16:0)	64 (20)	Not reversible & potentiation (44)+
(C 18:0)	68 (20)	" " " (48)
(C 19:0)	60 (18)	" " " (42)
(C 20:0)	68 (20)	" " " (48)
(C 22:0)	74 (20)	" " " (54)
(C 22:1)	70 (20)	" " " (50)
(C 20:4)	86 (18)	" " " (68)

*Data referred to as mean of two experiments.
§In brackets the percentage of platelet aggregability induced by ADP alone (0.6 μM).
+Potentiation calculated as difference in percent of aggregation.

Table II. Effect of Subcritical Concentration of FA in Thrombin-Induced (0.3 U) Platelet Aggregation

Subcritical concentration (0.4 mM) of FA with PRP on thrombin-induced (0.3 U) platelet aggregation	Thrombin-induced aggregation %	Behavior of aggregatory platelet response
(C 3:0)	24 (24)	Reversible
(C 4:0)	28 (26)	"
(C 5:0)	24 (22)	"
(C 8:0)	20 (20)	"
(C 10:0)	22 (20)	"
(C 11:0)	26 (26)	"
(C 14:0)	30 (26)	"
(C 16:0)	60 (24)	Not reversible & potentiation (36)[+]
(C 18:0)	60 (24)	" " (36)
(C 19:0)	80 (26)	" " (54)
(C 20:0)	60 (24)	" " (36)
(C 22:0)	88 (26)	" " (62)
(C 22:1)	80 (28)	" " (52)
(C 20:4)	96 (24)	" " (72)

Data referred to as mean of two experiments.
In brackets the percentage of platelet aggregability induced by thrombin alone (0.3 U).
[+]Potentiation calculated as difference in percent of aggregation.

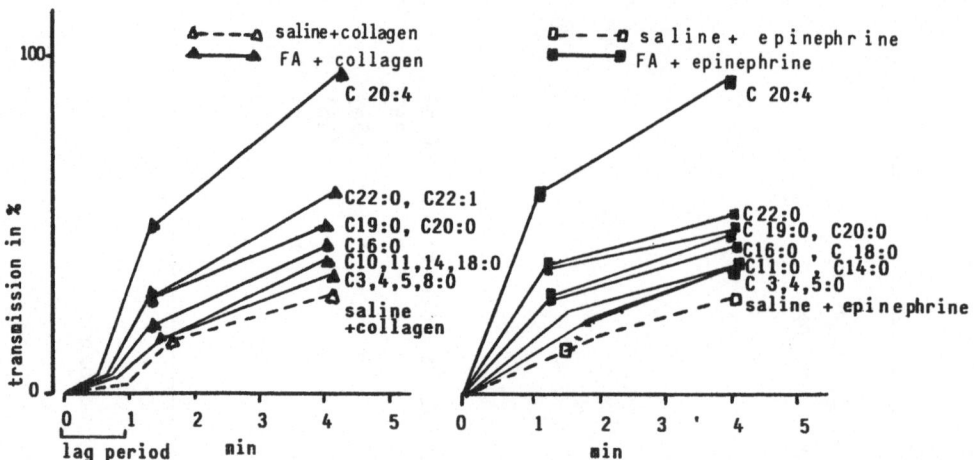

Fig. 1. Effect of subcritical concentration (0.4 mM) of FA in
         platelet aggregation induced by concentrations (1 µg)
         and epinephrine (1 µM).

DISCUSSION

     Our results showed that among several FA only those with a long
carbon chain, saturated or unsaturated, exerted a potentiating
effect on platelet aggregation induced by critical concentration
of ADP, epinephrine, collagen, and thrombin. With regard to the
mechanism whereby such potentiating effect could occur in our
experiments, it would be thought that exogenous FA added in vitro
might be incorporated in the platelet phospholipids fraction (20).
Subsequently, the exogenous FA, together with the endogenous ones
might be used in the prostaglandin biosynthesis through the cyclooxygenase pathway (19,21). The observation in our studies that AA,
a well known precursor of cyclic endoperoxides and thromboxane,
showed the greater potentiating effect on platelet aggregation
induced by critical concentrations of several agents would support
this view. However, it cannot be ruled out that long carbon chains
of FA, interacting with platelet bi-layering membranes (22), make
the platelets more sensitive to aggregating stimuli as already
observed for cholesterol and triglycerides (23). If this is true,
one could think that similar evidence could also occur in vivo
in those metabolic disorders which are associated with FA enhanced
plasma levels, thus resulting in a higher responsiveness of platelet
to minimal aggregating stimuli, which could represent a prominent
step towards thrombosis development.

REFERENCES

1. J.W. Kerr, I. MacAnlay, R. Pirrie, and B. Bronte-Stewart, Platelet aggregation by phospholipids and free fatty acids, Lancet, i:1296 (1965).
2. E.E. Nishizawa, T. Hovig, F. Lotz, H.C. Rowsell, and J.F. Mustard, Effect of a natural phosphatidyl serine fraction on blood coagulation, platelet aggregation, and haemostasis, Br. J. Haematol., 16:487 (1969).
3. J.R. Hampton and C.H. Bolton, Effects of phospholipids on platelet electrophoretic mobility, J. Atheroscl. Res., 9:131 (1969).
4. R.M. Jaffé, Influence of lipids on platelet function, in: "Platelet Function Testing," H.I. Day, ed., U.S. Dept. Health, Education, and Welfare, Philadelphia (1976).
5. M.J. Silver, J.B. Smith, C.M. Ingerman, and J.J. Kocsis, Arachidonic acid induced human platelet aggregation and prostaglandin formation, Prostaglandins, 4:863 (1973).
6. A.A. Mathé, P. Hedquist, K. Strandberg, and A.L. Crystal, Aspects of prostaglandin function in the lung (first of two parts), N. Engl. J. Med., 296:850 (1977).
7. R.L. Kinlough-Rathbone, H.J. Reimers, J.F. Mustard, and M.A. Packham, Sodium arachidonate can induce platelet shape change and aggregation which are independent of the release reaction, Science, 192:1011 (1976).
8. I.F. Charo, R.D. Feinman, T.C. Detwiler, J.B. Smith, C.M. Ingerman, and M.J. Silver, Prostaglandins, endoperoxides, and thromboxane A_2 can induce platelet aggregation in the absence of secretion, Nature, 269:66 (1977).
9. G. Hornstra and A. Vandelmans-Starrenburg, Induction of experimental arterial occlusive thrombi in rats, Atherosclerosis, 17:369 (1973).
10. J.B. Smith, M.J. Silver, and G.R. Webster, Phospholipase A_2 of human blood platelets, Biochem. J., 131:615 (1973).
11. N.W. Schoene and J.M. Iacono, Stimulation of platelet phospholipase A_2 activity by aggregating agents (abstract), Fed. Proceed., 34:257 (1975).
12. T.K. Bill and M.J. Silver, Phosphatidylcholine is the primary source of arachidonic acid utilized by prostaglandins synthetase (abstract), Fed. Proceed., 34:790 (1975).
13. D. Deykin, Emerging concept of platelet function, N. Engl. J. Med., 290:144 (1974).

14. H.J. Weiss, Platelet physiology and platelet function, <u>N. Engl.</u>
     <u>J. Med.</u>, 293:531 (1975).
15. J.M. Gerrard, J.C. White, and G.H.R. Rao, Labile aggregation
     stimulation substances (LASS): the factor from storage
     pool deficient platelets, <u>Br. J. Haematol.</u>, 29:657 (1975).
16. J.B. Smith, Platelet prostaglandin production and its impli-
     cations, <u>in</u>: "Prostaglandin and Thromboxane Research,"
     B. Samuelsson and R. Paoletti, eds., Raven Press, New
     York (1976).
17. G.V.R. Born, Aggregation of blood platelets by adenosine
     diphosphate and its reversal, <u>Nature</u>, 194:927 (1962).
18. N. Refsum, C. Lowery, and A. Nordøy, The effects of albumin-
     bound unsaturated fatty acids on platelets, <u>Haemostasis</u>,
     10:3 (1981).
19. C.G. Fenn and J.M. Littleton, Inhibition of platelet aggre-
     gation by ethanol <u>in vitro</u> shows specificity for aggre-
     gating agent used and is influenced by platelet lipid
     composition, <u>Thromb. Haemost.</u>, 48:49 (1982).
20. W. Siess, B. Scherer, B. Bohlig, P. Roth, I. Kurzmann, and P.C.
     Weber, Platelet-membrane fatty acid, platelet aggregation,
     and thromboxane formation during a mackerel diet, <u>Lancet</u>,
     i:441 (1980).
21. S.J. Shattil, R. Ahaya-Galindo, J. Bennet, R.W. Colman, and
     R.A. Cooper, Platelet hypersensitivity induced by
     cholesterol incorporation, <u>J. Clin. Invest.</u>, 55:115 (1975).
22. A.C.A. Carvalho and R.S. Lees, Platelets, intravascular
     coagulation, and fibrinolysis in hyperlipidaemias:
     relationship to thromboembolic complications, <u>Acta Med.</u>
     <u>Scand.</u>, (Suppl.), 642:101 (1980).

PROTEIN C - AN INHIBITOR OF BLOOD COAGULATION:

BIOCHEMISTRY, PHYSIOLOGY, CLINICAL ASPECTS

P.M. Mannucci, S. Viganò,
S. Antoncecchi, and N. Ciavarella

University of Milan
Hemophilia and Thrombosis Center A. Bianchi Bonomi &
University of Bari
Hemophilia Center

ISOLATION AND BIOCHEMICAL CHARACTERIZATION

In 1976, Stenflo (1) isolated a new vitamin K-dependent plasma glycoprotein from bovine plasma which he called protein C. Bovine protein C is composed of a light chain and a heavy chain held together by disulfide bonds. The amino-terminal region of the light chain contains γ-carboxyglutamic residues, necessary for Ca^{2+} binding, as do other vitamin K-dependent coagulation proteins. In 1979, Kisiel (2) isolated and partially characterized human protein C. He observed a single protein band after SDS-polyacrylamide gel electrophoresis with a mol. wt. of 62,000; after reduction with 2-mercaptoethanol, a heavy chain (mol. wt. 41,000) and a light chain (mol. wt. 21,000) were seen. Human and bovine protein C appear to be remarkably similar in terms of amino acid and carbohydrate composition with the exception of their histidine, valine, and N-acetylglucosamine contents. Human protein C appears to be more highly glycosylated than the bovine protein and contains 23% carbohydrate including mannose, galactose, glucosamine, and neuraminic acid. Aspartic acid is the amino-terminal residue of the heavy chain of human protein C. This is the same residue found as the amino-terminus of the heavy chain of bovine protein C. Despite this similarity, the next 17 residues from the amino-terminal

end of the human protein heavy chain are completely different from
those of bovine protein C. On the contrary, the amino acid sequence
for the light chain has great homology with that of bovine protein
C. Di Scipio and Davie (3) have recently reported that both human
and bovine protein C contain 10 γ-carboxyglutamic acid residues/
mole of protein. The γ-carboxyglutamic acids are in positions 6,
7, and 14 of the light chain.

In conclusion, protein C has the structure typical of other
vitamin K-dependent proteins, suitable for Ca^{2+} binding and for
acting on the surface of biological membranes.

ACTIVATION MECHANISM

Like other vitamin K-dependent coagulation proteins, protein
C exists in plasma in a zymogen form and is converted to a serine
protease by thrombin, Russell's viper venom, and trypsin. The enzyme
formed, referred to as activated protein C, has anticoagulant properties and is readily inactivated by diisopropyl fluorophosphate
(DFP). In the activation of bovine protein C, human thrombin cleaves
an Arg-Ile bond between residues 14 and 15 in the amino-terminal
region of the heavy chain of the protein, releasing a very small
activation peptide (mol. wt. 1,700). This cleavage results in the
formation of activated protein C with a new amino-terminal iso-
leucine residue in the heavy chain (4).

In the activation of human protein C by human thrombin, the
formation of serine protease activity coincides with the release
of a small peptide form the heavy chain. Most likely, the bond
cleaved during the activation of human protein C is the Arg-Leu
bond between residues 12 and 13 in the amino-terminal region of the
heavy chain. Cleavage of the dodecapeptide (mol. wt. 1,400) by
thrombin would then result in the formation of a new amino-terminal
sequence of Leu-Ile-Asp-Gly in the heavy chain of human activated
protein C. Thus, it seems highly probable that the basic mechanism
for proteolytic activation of human protein C is similar to that
observed for the activation of several other serine proteases,
including the vitamin K-dependent coagulation factors.

The rate of in vitro activation of purified protein C by
thrombin is too slow to have any physiological importance. However,
infusion of thrombin into the jugular vein of the dog leads to the

formation of anticoagulant activity in the circulation within 5 minutes after starting the infusion (5). The thrombin-catalyzed activation of protein C is markedly accelerated in vivo by an endothelial cell surface cofactor called thrombomodulin (6). Recently, thrombomodulin has been purified from rabbits' lungs (8). The activation of protein C by the thrombin-thrombomodulin complex is dependent on Ca^{2+} (6-8). When bound to thrombomodulin, thrombin cleaves fibrinogen and factor V poorly, if at all. The thrombomodulin-mediated inhibition of thrombin-catalyzed factor V cleavage is not dependent on protein C, since it also occurs in the absence of Ca^{2+} (9). The change in macromolecular substrate specificity of thrombin seems to be due to an alteration of some secondary binding sites or to a steric blockage of access of factor V and fibrinogen to the active site of thrombin, rather than to a change in the active site.

In conclusion, the thrombin-thrombomodulin complex seems to have two distinct anticoagulant activities: it leads to direct inhibition of thrombin procoagulant activity and activates protein C which can then function as a circulating anticoagulant.

ACTION MECHANISM AND PHYSIOLOGICAL REGULATION

The anticoagulant nature of activated bovine protein C was described over 20 years ago by Mammen et al. (10). They suggested that this potent anticoagulant activity, called autoprothrombin II-A, was derived from prothrombin that had been activated by limited thrombin proteolysis. A decade later, Marciniak (11) suggested that human autoprothrombin II-A was not derived from prothrombin, but was a distinct species-specific protein. In 1976, Stenflo (12) isolated an apparently unknown protein that Seegers et al. (13) showed to be immunologically indistinguishable from autoprothrombin II-A. The anticoagulant activity of activated bovine protein C was initially ascribed to a competitive inhibition of factor X_a (14,15). Subsequently, various investigators have shown that bovine activated protein C enzymatically inactivates bovine factors V and VIII by limited proteolysis (4,15,16,17). Walker has shown (18-21) that bovine protein S, another vitamin K-dependent protein, acts as a cofactor to potentiate the anticoagulant activity of activated bovine protein C and he ascribed to protein S the species-specificity reported for protein C (20).

Marlar et al. (22) have recently reported than when activated human protein C is incubated with normal plasma phospholipid and Ca^{2+}, factor V and VIII coagulant activities are rapidly destroyed. Activated protein C can inactivate activated forms of factors V and VIII more rapidly than non-activated factors. Activated protein C has no effect on either the contact system proteins (factors XI and XIII, prekallikrein, and high mol. wt. kininogen) nor the vitamin K-dependent coagulation factors (prothrombin, factors VII, IX, and X). No effect on fibrinogen was noted either.

In addition to its anticoagulant activity, activated protein C has been shown by Seegers et al. (23) and Zoltan and Seegers (24) to enhance fibrinolysis. Comp and Esmon (25) have confirmed that fibrinolytic activity can be generated both in vivo and in vitro in response to activated protein C. The in vivo administration of small quantities of activated protein C results in an increased rate of clot lysis. The active site of activated protein C is necessary for the enhancement of clot lysis because neither the DFP-inhibited enzyme nor the zymogen enhances clot lysis. This enhanced ability to lyse clot appears to result primarily from an elevation of circulating plasminogen activator levels. Despite this rise, no fibrinogenolysis occurs, as evidenced by normal levels of fibrinogen, and there is no elevation of fibrinogen degradation production during prolonged infusion of activated protein C. This suggests that the role of the enzyme in clot lysis is different from those of urokinase or streptokinase which activate circulating plasminogen to plasmin with subsequent proteolysis of fibrinogen. Perfusion of isolated dog tissues does not release plasminogen activator from the vasculature.

However, when activated protein C is added to blood in vitro and subsequently neutralized with specific antibodies, the rate of clot lysis is increased when this blood is then reinjected into the animal. The addition of activated protein C to plasma alone is sufficient to generate the fibrinolytic activity, but the effect is more marked when the enzyme is added to a mixture of plasma and blood cells. Comp and Esmon (25) have suggested that activated protein C might generate a secondary messenger which in turn causes the increase in circulating plasminogen activator activity.

Canfield and Kisiel (26) have recently succeeded in purifying to homogeneity an activated protein C-binding protein from normal

human plasma. This binding protein (mol. wt. 54,000) is a glycoprotein and possesses an amino-terminal sequence of Gly-Arg-Thr-
Cys-Pro-Lys-Pro-Asp which has considerable homology with bovine
cholostrum inhibitor and pancreatic trypsin inhibitor, but no apparent sequence homology with the known plasma serine protease
inhibitors. Affinity-purified antibody against this binding protein
immunoprecipitated a complex of radiolabeled and native-binding
proteins from normal human plasma. There was essentially no complex
formation in plasma immunodepleted of the binding protein.

In conclusion, protein C has an important physiological function in hemostasis, in modulating clot formation, and in clot lysis.
The effects of the enzyme are regulated, in turn, by a naturally
occurring inhibitor.

CLINICAL SIGNIFICANCE

The physiological role of protein C is now less obscure and
its clinical importance is becoming increasingly evident. Marlar
and Griffin in 1980 (27) reported that protein C inhibitory activity was detectable in plasma. This inhibitory activity could not
be seen in plasmas from four unrelated patients with combined factor V/VIII deficiency, but it was present in normal amounts in
patients with isolated factor V or factor VIII deficiencies. Since
activated protein C readily destroys factor V and factor VIII
coagulant activities (18), they suggested that the molecular basis
for combined factor V/VIII deficiency is the congenital lack of
this inhibitor. Canfield and Kisiel (26) have recently confirmed
that normal human plasma contains an inhibitor directed against
activated protein C. However, at variance with the findings of
Marlar and Griffin, they found that plasma from four patients with
combined factor V/VIII deficiency inhibited the anticoagulant and
amidolytic activities of activated human protein C at essentially
the same rate as normal pooled plasma. Furthermore, incubation of
radiolabeled activated protein C, with either normal or combined
factor V/VIII-deficient plasma, resulted in complex formation at
the same rate and to the same extent. Finally, quantitative electroimmunoassay indicated equal levels of activated protein C binding
protein in normal plasma and in plasma from the four patients with
combined factor V/VIII deficiency. Therefore, the role of activated
protein C and its naturally occurring inhibitor in the pathophysiology of combined factor V/VIII deficiency is still unsettled.

Since protein C is a potent anticoagulant and profibrinolytic agent, one could hypothesize that an inherited deficiency of protein C would cause a thrombophilic state. Until now, inherited thrombophilia has been associated with abnormalities of three proteins: antithrombin III, fibrinogen, and plasminogen. The availability of specific antisera against protein C has led to the development of electroimmuno assays which have clarified the clinical significance of the inhibitor. In 1981, Griffin et al. (28) reported a family with a history of recurrent venous thromboembolism due to an inherited deficiency of protein C. The propositus, his father, and his paternal uncle had 38-49% of normal levels of protein C antigen, whereas, unaffected family members had normal levels. Since no assay of functional protein C in plasma was available, the study of protein C was limited to immunological assay. In 1982, Bertina et al. (29) described a similar Dutch family with a history of thrombotic disease associated with an isolated deficiency of protein C antigen. In 1982, Mannucci and Viganò (30) reported data about acquired defects of protein C. In healthy controls, there were no differences in protein C related to age or sex. The low levels found in 12 healthy full-term newborn infants wre probably a reflection of liver immaturity at birth, also expressed in the low levels of vitamin K-dependent and contact-phase clotting factors. The protein C levels in 20 women in the last trimester of normal pregnancy were no different from those in healthy non-pregnant women. Protein C concentrations were low in chronic liver disease, in degrees roughly proportional to the severity of the disease and thus to the impairment of protein synthesis, suggesting that the liver is probably the site of synthesis of this protein.

However, one cannot rule out the possibility that low plasma concentrations of protein C are due to increased turnover, as has previously been found for other hemostatic components in chronic liver disease. In clinical conditions associated with disseminated intravascular coagulation (DIC), protein C levels were very low and sometimes unmeasurable (30). The most likely explanation of this being that protein C activated by thrombin during DIC is rapidly cleared from the circulation. Similar results were obtained by Griffin et al. (28) for patients with DIC. It is possible that during DIC, anoxia or endotoxin (or both) could damage endothelial cells allowing the endothelial-cell cofactor to be exposed in greater than normal quantities; hence, protein C activation would be much increased and its clearance from plasma accelerated. This hypothesis is supported by the finding of very low or unmeasurable levels of protein C in adult respiratory distress syndrome, a condition characterized by extensive pulmonary endothelial damage.

The period immediately after minor or major surgical operation was also associated with an acquired defect of protein C (30). A possible explanation is that during surgery and immediately afterwards tissue damage induced an in vivo activation of blood clotting, resulting in increased thrombin formation. In conjuction with the endothelial-cell cofactor, thrombin might in turn activate protein C, leading to faster removal of protein C from the circulation. Increased levels of protein C have also been found in patients with ischemic cardiovascular disease, particularly in those with active disease as expressed by increased fibrinopeptide A plasma levels (31). It is possible that thrombin generation induces an increased turnover of the protein, leading to a faster rate of synthesis of protein C. Studies are currently in progress in our laboratory to ascertain the behavior of protein C in other acquired conditions, such as ischemic cerebrovascular disease and acute leukemias.

In conclusion, the recent availability of a simple and reproducible electroimmune assay for protein C antigen has allowed a considerable amount of information about the pathological role of this new vitamin K-dependent protein to be obtained. However, the lack of a similarly reliable biological assay hampers drawing firm conclusions.

REFERENCES

1.   J. Stenflo, A new vitamin K-dependent protein, J. Biol. Chem., 251:355-363 (1976).

2.   W. Kisiel, Human plasma protein C, J. Clin. Invest., 74:761-769 (1979).

3.   R.G. Discipio and E.W. Davie, Characterization of protein S a γ-carboxyglutamic acid containing protein from bovine and human plasma, Biochemistry, 18:899-904 (1979).

4.   W. Kisiel, W.M. Canfield, L.H. Ericsson, and E.W. Davie, Anticoagulant properties of bovine plasma protein C following activation by thrombin, Biochemistry, 16:5824-5831 (1977).

5.   P.C. Comp, R.M. Jacocks, G.L. Ferrell, and C.T. Esmon, Activation of protein C in vivo, J. Clin. Invest., 70:127-134 (1982).

6. W. Owen and C.T. Esmon, Functional properties of an endothelial
    cell cofactor for thrombin-catalyzed activation of
    protein C, J. Biol. Chem., 256:5532-5535 (1981).

7. C.T. Esmon and W.G. Owen, Identification of an endothelial
    cell cofactor for thrombin-catalyzed activation of protein
    C, Proc. Natl. Acad. Sci. USA, 78:2249-2252 (1981).

8. N.L. Esmon, W.G. Owen, and C.T. Esmon, Isolation of a membrane
    bound cofactor for thrombin-catalyzed activation of
    protein C, J. Biol. Chem., 257:859-864 (1982).

9. C.T. Esmon, N.L. Esmon, and K.W. Harris, Complex formation
    between thrombin and thrombomodulin inhibits both thrombin-
    catalyzed fibrin formation and factor V activation, J.
    Biol. Chem., 257:7944-7947 (1982).

10. E. Mammen, W. Thomas, and W.H. Seegers, Activation of purified
    prothrombin to autoprothrombin I or autoprothrombin II
    (platelet cofactor II) or autoprothrombin II A, Thromb.
    Diath. Haemorrh., 5:218-250 (1960).

11. E. Marciniak, Inhibitor of human blood coagulation elicited
    by thrombin, J. Lab. Clin. Med., 79:924-934 (1972).

12. J. Stenflo, A new vitamin K-dependent protein: purification
    from bovine plasma and preliminary characterization,
    J. Biol. Chem., 251:355-363 (1976).

13. W.H. Seegers, E. Novoa, R. Henry, and H. Hassouna, Relationship
    of "new" vitamin K-dependent protein C and "old" autopro-
    thrombin IIA, Thromb. Res., 8:543-553 (1976).

14. G. Murano, W.H. Seegers, and R. Zolton, Autoprothrombin IIA:
    a competitive inhibitor of autoprothrombin C. A review
    with additions, Thromb. Diath. Haemorrh., (Suppl.) 57:
    305-314 (1974).

15. W.H. Seegers, R.A. Marlar, and D. Wall, Anticoagulant effects
    of autoprothrombin IIA and prothrombin fragment 1, Thromb.
    Res., 13:233-243 (1978).

16. C. Esmon, P. Comp, and F. Walker, Functions for protein C,
    in: "Vitamin K Metabolism and Vitamin K-Dependent Proteins,"
    J. Suttie, ed., University Park Press, Baltimore,
    (1980).

17. G. Vehar and E. Davie, Preparation and properties of bovine
    factor VIII (Antihemophilic factor), Biochemistry, 19:
    401-410 (1980).

18. F.J. Walker, P.W. Sexton, and C.T. Esmon, The inhibition of
    blood coagulation by activated protein C through the
    selective inactivation of activated factor V, Bioch. et
    Bioph. Acta, 571:333-342 (1979).

19. F.J. Walker, Regulation of activated protein C by a new protein, J. Biol. Chem., 255:5521-5524 (1980).

20. F.J. Walker, Regulation of bovine activated protein C by protein S: the role of the cofactor protein in species specificity, Thromb. Res., 22:321-327 (1981).

21. F.J. Walker, Regulation of activated protein C by protein S. The role of phospholipid in factor Va inactivation, J. Biol. Chem., 11128-11131 (1981).

22. R.A. Marlar, A.J. Kleiss, and H.J. Griffin, Mechanism of action of human activated protein C, a thrombin-dependent anticoagulant enzyme, Blood, 59:1067-1072 (1982).

23. W.H. Seegers, C. Mc Coy, H. Groben, N. Sakuragaza, and B. Agrawal, Purification and some properties of autoprothrombin IIA: an anticoagulant perhaps related to fibrinolysis, Thromb. Res., 1:443-460 (1972).

24. R.P. Zoltan and W.H. Seegers, Autoprothrombin IIA: thrombin removal and mechanism of induction of fibrinolysis, Thromb. Res., 3:23-33 (1973).

25. P.C. Comp and C.T. Esmon, Generation of fibrinolytic activity by infustion of activated protein C into dogs, J. Clin. Invest., 68:1221-1228 (1981).

26. W.N. Canfield and W. Kiesel, Evidence of normal functional levels of activated protein C inhibitor in combined factor V/VIII deficiency disease, J. Clin. Invest., 70: 1260-1272 (1982).

27. R.A. Marlar and J.H. Griffin, Deficiency of protein C inhibitor in combined factor V/VIII deficiency disease, J. Clin. Invest., 66:1186-1189 (1980).

28. J.H. Griffin, B. Evatt, T.S. Zimmerman, A.J. Kleiss, and C. Wideman, Deficiency of protein C in congenital thrombotic disease, J. Clin. Invest., 68:1370-1373 (1981).

29. R.M. Bertina, A.W. Broekmans, I.K. Van der Linden, and K. Nerkens, Protein C deficiency in a Dutch family with thrombotic disease, Thromb. Haemost., 48(1):1-5 (1982).

30. P.M. Mannucci and S. Viganò, Deficiencies of protein C, an inhibitor of blood coagulation, Lancet, 2:463-467 (1982).

31. S. Favilla, R. Abbate, C. Rostagno, G.F. Gensini, and P.M. Mannucci, Modificazioni della proteina C nella cardiopatia ischemica e suoi rapporti con la concentrazione plasmatica del fibrinopeptide A, 7° Congr. Naz. SISET, Ist. Sup. Sanità Roma, Abstract Book, p. 71 (1982).

IMMUNE COMPLEXES AND THROMBOSIS

G. Camussi and F. Gavosto

Clinica Medica I
Via Genova 3
Turin, Italy

INTRODUCTION

The antigen-antibody reaction can activate several humoral
and cellular systems that exist in the circulation as precursors
or in an inactive state and need to be triggered or activate in
order to participate in the inflammatory reaction. While in the fluid
phase, biological activities generated from activated molecules or
reactive cells are modulated by several serum inhibitors, activation
appears to be fully operating on "surfaces" such as connective
tissue elements, basement membranes, cells or immune complexes (IC).
"Surface" activation favors changes of spatial orientation of the
native molecules that are necessary to trigger the entire sequence
in humoral system or various functions in cells. IC may activate
coagulation: 1) by activation of Hageman Factor (HF) through a
mechanism requiring an intact complement system, or as result of
tissue damage and collagen exposure, or inflammatory cell activation, 2) by stimulation of platelets either directly through their
membrane receptors or through secondary mediators released from
activated leukocytes.

The HF system not only initiates the coagulation cascade, but
also represents one of the major systems in plasma possessing
inflammatory potential. HF initiates the sequential activation of
factor XI, IX, VIII, and X. Platelets may contribute by accelerating
this reaction ("intrinsic clotting mechanism"). However, the

"extrinsic clotting mechanism" can also be involved by IC consequent
to factor X activation by thromboplastin released from damaged
tissue. Through a complex series of interactions, HF also activates
kinin as well as the fibrinolytic and complement systems. Apart
from being an integral part of the hemostatic and clotting mechanisms,
platelets play an important role in the inflammatory reactions
induced by immunological mechanisms. When antigen-antibody complexes
are formed in vivo, platelets may be instrumental both in enhancing
vascular permeability leading to IC localization in the vasculature
and in maintaining the inflammatory reaction. As for PMN, these
biological effects are mediated by lysosomal enzymes that are
released from stimulated platelets.

THE CONTACT (HAGEMAN FACTOR) SYSTEM ACTIVATION

     Early data suggested that IC were capable of activating the
HF system, but primary and secondary events could not be distinguished (1). Subsequent studies indicated that IC do not directly
initiate the coagulation cascade (2), and the following secondary
mechanisms were implicated:

1) Complement-Mediated Activation A variety of complement-activating
substances, including IC, were found to initiate blood coagulation
through a complement-mediated pathway (3), and were implicated in
intravascular blood coagulation. The clot-promoting activity
of IC, as for endotoxins, requires an intact complement system,
as demonstrated by an almost complete lack of effect of IC on blood
from rabbits with an inherited complement deficiency (3). A role
for C6 and C7 was suggested, even though the precise mechanism of
interaction between the complement system and blood coagulation
remains to be defined. The activation of the alternative pathway
of complement appears to be effective while the presence of C1, C2,
and C4 was not necessary (3).

2) Exposure of Collagen Secondary to Vascular Injury An indirect
activation may possibly take place after exposure of vascular
basement membranes (mucopolysaccharides and collagen) as result
to tissue injury consequent to IC deposition.

3) Cleavage and Activation of HF by Cellular Enzymes It was shown
that enzymes released from cells upon stimulation cleave and
activate surface-bound HF (4). Peripheral leukocytes obtained from

allergic subjects are able to release a kallikrein-like enzyme when
challenged by specific antigen through an IgE-mediated mechanism
(5). The released enzymes rapidly cleave the surface-bound HF,
acting both at site 1 and particularly at site 2 of this molecule
with consequent activation (3). Similarly, other cells that may be
stimulated by IC such as polymorphonuclear neutrophils (PMN) and
macrophages were able to release proteases capable of activating HF.

Contact between plasma and a negatively charged surface
generates activity of the system. Three molecules are mainly
involved: HF (clotting factor XII), prekallikrein, and high molecular
weight (HMW) kininogen (see Ref. 4). The combination of HF
and prekallikrein results in a reciprocal activation of the two
molecules which, in turn, sets in motion the system. Furthermore,
other enzymes, released from activated cells, may activate HF in
a way quite similar to kallikrein (4). HF, in its activated form,
is able to further activate prekallikrein or clotting factor XI, or
the intrinsic clotting system and factor VII of the extrinsic
clotting system. Moreover, kallikrein, an endopeptidase, can also
cleave bradikinin from HMW kininogen and activate plasminogen. The
molecules of this system possess a broad range of biological activities that are prominent in inflammation (4). Among these molecules,
bradikinin increases vascular permeability, induces hypotension
and smooth muscle contraction, evokes pain in tissue pre-treated
with prostaglandins, and favors the generation of arachidonate-
derived metabolites. Activated HF causes an increased vascular
permeability in the skin. Plasmin activates C1 and C3 fractions
of the complement system. Plasmin, therefore, can activate the
complement cascade by the classical and alternative pathways, thus
amplifying the IC-induced activation of this system. Plasmin also
activates the fibrinolytic system as it generates both fibrinogen
and fibrin peptides that are able to increase vascular permeability
and to elicit the chemotactic response to PMN. The activity of
both HF and kallikrein generated by the contact system on "surfaces"
is given off into the fluid phase. This phenomenon in biological
systems may be of some importance in the dissemination of activity
(4).

PLATELET ACTIVATION

Immunologic reactions involving platelets can be categorized
into three types: 1) IC-induced reactions which may occur either

as a) direct IC-platelet interaction or b) indirectly through the
mediation of the complement system, 2) leukocyte-mediated platelet
activation via secondary mediators, 3) delayed hypersensitivity or
cell-mediated immunity.

1a) IC-Platelet Interaction IC trigger serotonin release from
human platelet-rich plasma and washed human platelets (6). In
order for platelets to be activated by IgG, bridging of some cell
surface molecules may be the necessary signal. Considerable
attention has been paid to the possibility that interaction of a
specific cell-surface molecule occurs with a specific portion of
the IgG. Aggregated IgG of all 4 subclasses stimulate platelets
while other immunoglobulins do not (7). This suggests that the
critical site(s) are not on the light chains. Accordingly F(ab')
2-IC (7) and aggregated F(ab')2 (8) are incapable of initiating
the secretion of serotonin from human platelets. In contrast,
aggregated Fc fragments not only lead to serotonin secretion, but
are able to block platelet stimulation by IC (8). The existence
of platelet Fc receptors has been evaluated in studies on the
binding of the relevant ligand to the platelets. Using chemically
aggregated IgG, it was seen that binding to human platelets was
tight and secretion was triggered when 40-70 aggregates/cell
were bound (9,10). However, although these studies support the
role of the Fc region of IgG in IC-induced platelet stimulation,
no consistent evidence of saturable binding to a limited number of
sites/platelets is available, thus ruling out the hypothesis of a
unique platelet receptor for Fc. A membrane protein with a molecular
weight of 250,000, located in the plasma membrane protein array,
binds Fc fragments as well as aggregated Fc, suggesting that this
protein may at least in part make up the membrane Fc receptor (11).
However, at the present time, nothing is known about the surface
localization of this protein as well as its role in IC-induced
secretion. Multivalent IC may trigger platelets by cross-linking
membrane proteins that could eventually fix receptors in close
proximity one to each other and markedly limit both the rotational
and transitional diffusion of the receptors. Further studies on
the Fc receptors have led to the suggestion that C1 of the complement
system or a C1-like molecule may actually be the platelet Fc
receptor (12). In fact, there exists a rough correlation between
complement fixing ability and platelet stimulating activity, and
the ability of C1 to block IgG aggregates-induced platelet stimulation. In addition, small quantities of C1 hemolytic activity were

reported on platelets. Recent data on the role of platelet-bound
Cl hemolytic activity in platelet secretion to IgG aggregates
indicate that Cl-platelet interaction is a sufficiently low affinity
one, thus allowing Cl activity to be removed during a single gel
filtration (13). However, small quantities of Cl or Clg, which are
not hemolytically active, may in fact represent the platelet Fc
receptor. In contrast, the lipid A constituent of the bactericidal
polysaccharides (14) as well as carbohydrate-modified IgG aggregates
(15) activate Cl, but do not trigger platelet secretion. Furthermore, protease-treated platelets are still capable of binding IgG
aggregates and release serotonin, even though, in this condition,
one would expect Cl or its subcomponent to be cleaved (10).
Structural studies (11) provided no evident similarities of the Fc-
binding protein with any Cl subcomponent. Surface-bound IgG is,
therefore, a platelet stimulus. Platelet stimulation may occur in
situations of IgG bound to an apposing cell (13). This may be the
case because it is supported by the failure of surface-bound IgG to
stimulate Fc receptor-negative rabbit platelets. Anti-platelet
antibodies may stimulate human platelets by a mechanism involving
stimulation of Fc receptors on apposing cell (13). The constituents
of microorganisms are major contributors to IC formation. These
types of IC are likely to encounter the platelets. Thus, the studies
emphasizing platelet stimulation by IC of this type have an important
bearing on immunopathology. Bacterial products may not only serve
as antigens, but may cause, per se, immunologically non-specific
IgG-dependent stimulation of platelets (13). Furthermore, they may
markedly potentiate platelet stimulation by IgG-containing IC.
When platelet interaction with yeast cell walls (zymosan) was studied
in plasma, platelet aggregation and secretion occurred following a
long-lag phase (16). The requirement for fibrinogen (16) and C5-C7
but not C8-C9 complement fractions (17) was shown. Clear demonstration was given that platelet stimulation by plasma-coated zymosan
is based on zymosan anti-zymosan interaction, since a platelet Fc
is critical for zymosan-induced platelet stimulation (18). Serum
decomplemented plasma markedly reduced the effect of zymosan on
platelets. This effect may be attributed to Cl, IgG monomer, and/or
other unidentified plasma component(s) interfering with IC-platelet
interaction. A second mechanism of enhancement of the platelet
response is seen in the presence of bacterial lipopolysaccharides
(LPS)-containing-IC (13). The lipid A region of the LPS molecules
is present in different quantities in different strands of bacteria.
Lipid A, in its isolated form, was shown to enhance platelet

serotonin secretion in response to IC approximately 50-fold (14).
In evaluating potential mechanisms of this phenomenon, an attractive
hypothesis considers the formation of a ternary complex between
lipid A-immune aggregate and platelet surface (13). Lipid A could
form a complex with IgG-containing immune aggregates and may per
se bind at the same time to the platelet surface. Addition of
Lipid A to an immune aggregate also increases its affinity for the
platelet surface providing an additional site to the attachment of
the platelets (13). An additional mechanism has been recently
described (19). Staphylicoccal protein A has a high affinity for
the Fc fragment of IgG 1, 2, and 4 monomers. IC of this type
stimulate human platelet secretion independently from complement
even in the presence of plasma.

1b) Platelet-Complement System Interaction The interaction between
platelets and the complement system includes: 1) activation of
platelets by complement components and, 2) activation of complement
components by platelets. Rabbits platelets are involved by IC
either directly through the C3b receptors or as innocent bystanders,
leading to complement-dependent lysis of platelets (for review
see Ref. 13). In contrast, human platelets do not interact with
IC through C3b receptors and do not undergo a lytic reaction of the
innocent bystander type due to the absence of C3b receptors on
human platelets (13). However, as mentioned above, zymosan-induced
human platelet stimulation requires the alternative pathway of
complement activation (through at least C3) as well as fibrinogen
(16,17). The question as to what effect the complement system may
have on C3b receptor-negative platelets remains an intriguing one.
Recent studies showed that arachidonate-mediated release of C^{14}-
serotonin and thromboxane B_2 (TXB_2) was significantly enhanced in
the presence of complement (20). Only the complement components
C3-C5-C6-C7-C8-C9 were required for this activity in the absence of
any other mechanism known to activate the classical and alternative
pathways. The complement-mediated effects on human platelets could
be inhibited by aspirin and indomethacin, both inhibitors of platelet
cyclooxygenase. The possibility arises that C3 in association with
C5 and C9 is required for mobilization of the arachidonic acid
from the phospholipid of the platelet membrane. Once arachidonic
acid is mobilized, C3 is no longer required, since C5-C9 are sufficient to modulate this pathway leading to enhanced production of
TXB_2. The binding of thrombin to the surface of human platelets
in the presence of complement components leads to the generation

of a C5-convertase enzyme, probably a thrombin-C3b complex. This
enzyme has been reported to be similar to both alternative and
classical C5a convertases. An assembly process could be initiated
by the thrombin-C3b complex on the platelet surface with the
subsequent deposition of dimers of C5-C9 complex and the appearance
of ultrastructural lesions (20). These data are relevant not only
because they support the hypothesis of a new pathway of complement
activation, but also because they suggest the enhancing role of
complement in thrombin-induced platelet activation. Complement
activation may also occur as the consequence of the release of
proteases (elastase-like enzymes) from aggregating platelets
(reviewed in Ref. 13). These proteases are able to cleave native
Synacthen, the specific antigen, or C5a anaphylatoxin (31). On the
in some cases complement components such as C1g bound to IC or to
IgG aggregates may be inhibitory of platelet stimulation via the
Fc receptors (12). This effect presumably operates by steric
hindrance and may have a modulatory role in IC-platelet interaction
in plasma.

2) Leukocyte-Mediated Platelet Interaction Platelets may be involved
secondarily by leukocyte-dependent mechanisms. IC-stimulated
leukocytes release lipid mediators such as the metabolites of
arachidonate (prostaglandings, thromboxanes, and leukotrienes) or
a platelet-activating factor (PAF). Recently, PAF has been identified as a 1-O-octadecyl/hexadecyl-2-acetyl-sn-glyceryl-3-phospho-
rylcholine (21,22). Henson first described the occurrence of a
"soluble" factor that led to the release of histamine from platelets
(23). This factor that was recovered from the supernatants of
antigen-stimulated sensitized leukocytes was considered responsible
for leukocyte-platelet communication. This was confirmed in studies
by Siraganian and Osler (24) and by Benveniste, Henson, and
Cochrane (25). The latter authors indicated the rabbit basophil
as a likely cellular source for the newly-discovered mediator in
the fluid phase, which they termed PAF after its ability to aggregate platelets and to release their granular constituents. An
in vivo release of PAF has been shown in experimentally-induced
anaphylaxis (26) and in immune complex pathology in the rabbit
(27,28). Studies conducted on human mixed leukocyte preparations
showed the release of PAF after stimulation with a specific
antigen, C5a anaphylatoxin and immune complexes (IC) (29). The
possibility that human basophils can release PAF was inferred from
the concurrency between PAF release and basophil degranulation.

Lewis et al. (30) reported the release of PAF-like material from
purified basophilic leukemic cells. Other investigators, however,
could not show an IgE-dependent release of PAF from human basophils.
This controversy has been settled by the increasing recognition
that a variety of human cell types are able to release PAF. Nevertheless, the specificity seems to reside more in the PAF-releasing
stimulus rather than in the cell type (31). Evidence exists that
human basophils in basophil-rich buffy coat cells are also implicated
in PAF release, based on the specificity of Synacthen in de-
granulating basophils concomitantly with PAF and histamine release,
and on electron microscopy studies showing the formation of aggregates of degranulated platelets around basophils stimulated with
Synacthen, the specific antigen or C5a anaphylatoxin (31). On the
contrary, no doubt existed on the capacity of PMN and monocytes to
generate and release PAF, given the high degree of purity of both
PMN and monocyte preparations (31,32).

Purified human PMN release PAF during phagocytosis of C3b-
opsonized or IgG-coated baker's yeast spores or when stimulated with
C5a, cationic proteins (CP) from neutrophils and their desarginated
cleavage fragments, C5a des Arg and CP des Arg (31). When C5a-
stimulated or phagocytic PMN are placed in the presence of platelets,
morphologic evidence of PMN-aggregated platelet interaction is seen
despite complete blockade of adenosine diphosphate (ADP), and
cyclooxygenase-metabolite-dependent pathways of aggregation (31).
A positive correlation is found between the amount of PAF release
and the number of eosinophils present in eosinophil-enriched
preparations. We obtained ultrastructural evidence of phagocytic
eosinophil aggregated platelet interaction. Human monocytes release
PAF only when adherent. The adherence to plastic surface of monocytes not only accelerates the release of PAF, but also increases
the amount of releasable PAF and the number of available surface
receptors for Fc, C3b, and C3d. A mechanism specifically related
to surface Fc, C3b, and C3d membrane receptors accounts for the
effect of the phagocytic stimuli. Thus, the release of PAF occurs
during phagocytosis of opsonized baker's yeast spores or immune
complexes. C5a, a soluble agent, which induces a receptor-mediated
release of PAF from PMN, fails to induce PAF release from monocytes.
Monocytes generate higher amounts of PAF than PMN. Recently, we
have demonstrated that endothelial cells are able, in certain in
vitro and in vivo conditions of stimulation, to generate and release
PAF. PAF is the most potent platelet activating agent so far

described as it induces aggregation, release of alpha-granules and dense bodies constituents, activation of membrane phospholipids metabolism, and phosphorylation of membrane protein at nanomolar concentrations (for review see Ref. 33). There are large differences in sensitivity to PAF among various species. Thus, amounts of PAF needed to induce a 50% activation (aggregation or serotonin release from washed platelets) are $1x10^{-10}$ M, $2x10^{-9}$ M, and $1x10^{-9}$ M for rabbit, guinea pig, and man respectively. PAF induces a dose-dependent activation of human platelets both in plasma (PRP) or washed in the presence of fibrinogen (34,37). $1x10^{-7}$ M or $1x10^{-8}$ M of PAF induce irreversible aggregation in PRP or of washed platelets respectively, and reversible aggregation is obtained at lower concentrations. Aggregation starts 2-3 seconds after addition of PAF and reaches the peak after 60-90 seconds, while the release reaction is maximal within 20 seconds and appears independent from stirring which is required for aggregation to take place. Several studies indicated that PAF-induced platelet activation is independent from the release of endogenous ADP and of the cyclooxygenase-derived metabolites (34,37). However, studies with inhibitors of the lipooxygenase indicated that the metabolites of arachidonic acid generated through this pathway may have a role in PAF-induced human platelet aggregation.

3) Platelets in Cell-Mediated Immune Reactions Several reports indicate that simulated lymphocytes produce lymphokines with a platelet-stimulating effect (for review see ref. 13). In addition, macrophages can synthetize PAF and platelet-reactive arachidonate derivatives that may cooperate in platelet activation. This area requires, however, further investigation.

REFERENCES

1. V. Eisen and H.G. Smith, Plasma kinin formation by complexes of aggregated γ globlulins and serum proteins. Br. J. Exp. Pathol., 51:328 (1970).

2. C.G. Cochrane, K.D. Wuepper, B.S. Aiken, S.D. Revak, and H.L. Spiegelberg, The interaction of Hageman factor and immune complexes, J. Clin. Invest., 51:2736 (1972).

3. T. Zimmerman and H.J. Mueller-Eberhard, Blood coagulation initiation by a complement-mediated pathway, J. Exp. Med., 134:1601 (1971).

4. C.G. Cochrane, S.R. Revak, R.C. Wiggins, and J.H. Griffin, The Hageman factor system in inflammation, in: "Advances

in Inflammation Research," G. Weissman et al., eds., Raven Press, New York, p. 249 (1979).

5.   H.H. Newball, R. Talamo, and L.M. Lichtenstein, Release of leukocyte kallikrein mediated by IgE, Nature, 254:635 (1975).

6.   J.H. Humphrey and R. Jaques, The release of histamine and 5-hydroxytriptamine (serotonin) from platelets by antigen-antibody reactions (in vitro), J. Physiol., 128:9 (1955).

7.   P.M. Henson and H.L. Spiegelberg, Release of serotonin from human platelets induced by aggregated immunoglobulins of different classes and subclasses, J. Clin. Invest., 52: 1282 (1973).

8.   D. Israels, G. Nisli, F. Paraskevas, and C.G. Israels, Platelet Fc receptor as a mechanism for antigen-antibody complex induced platelet injury, Thromb. Diath. Haemorrh., 29:434 (1973).

9.   S.L. Pfueller, S. Weber, and E.F. Luescher, Studies on the mechanisms of the human platelet release reaction induced by immunologic stimuli. III. Relationship between binding of soluble IgG aggregates to the Fc receptor and cell responses in the presence and absence of plasma, J. Immunol., 118:514 (1977).

10. S.L. Pfueller, C.S.P. Jenkins, and E.F. Luescher, A comparative study of the effect of modification of the surface of human platelets on the receptors for aggregated immuno-globulins and for ristocetin-von Willebrand factor, Biochem. Biophys. Acta, 465:614 (1977).

11. C.M. Cheng and J. Hawiger, Affinity isolation and characterization of immunoglobulin G Fc fragment-binding glycoprotein from human blood platelets, J. Biol. Chem., 254:2165 (1979).

12. S.L. Pfueller and E.F. Luescher, The effects of aggregated immunoglobulins on human blood platelets in relation to their complement-fixing abilities, J. Immunol., 109:517 (1972).

13. P.M. Henson and M.H. Ginsberg, Immunological reaction of platelets, in: "Platelets in Biology and Pathology," Gordon, ed., North-Holland Biochemical Press, p. 265, (1981).

14. M.H. Ginsberg and P.M. Henson, Enhancement of platelet response to immune complexes and IgG aggregates by lipid A-rich bacterial polysaccharides, J. Exp. Med., 147:207 (1978).

15. S.L. Pfueller and E.F. Luescher, The effects of aggregated immunoglobulins on human blood platelets in relation to their complement-fixing abilities. I. Studies of immuno-globulins of different types, J. Immunol., 109:526 (1972).

16. S.L. Pfueller and E.F. Luescher, Studies of the mechanisms of the human platelet release reaction induced by immunologic stimuli. II. The effects of zymosan, J. Immunol., 112:1211 (1974).

17. R.T. Breckenridge, S.T. Rosenfeld, K.S. Graff, and J.P. Leddy, Hereditary C5 deficiency in man. III. Studies of homeo-stasis and platelet response to zymosan, J. Immunol., 118:12 (1977).

18. S.E. Martin, R.T. Breckenridge, S.I. Rosenfeld, and J.P. Leddy, Responses of human platelets to immunologic stimuli: independent roles for complement and IgG in zymosan activation, J. Immunol., 120:9 (1978).

19. J. Hawiger, S. Steckley, D. Hammond, C. Cheng, S. Timmons, A.D. Glick, and R.M. des Prez, Staphylococci-induced human platelet injury mediated by protein A and immuno-globulin G Fc fragment receptor, J. Clin. Invest., 64:931 (1979).

20. N.J. Polley, R.L. Nachman, and B.B. Weksler, Human complement in the arachidonic acid transformation pathway in platelets, J. Exp. Med., 153:257 (1981).

21. J. Benveniste, M. Tencé, P. Varenne, J. Bidault, C. Boullet, and J. Polonsky, Semi-synthèse et structure proposée du facteur activant les plaquettes (PAF): PAF-acether, un alkyl ether analogue de la lysophosphatidylcholine, C.R. Acad. Sci. (Paris), 289:1037 (1979).

22. C.A. Demopoulos, R.N. Pinckard, and D.J. Hanahan, Platelet-activating factor. Evidence for 1-O-alkyl-sn-glyceril-3-phosphorylcholine as the active component (a new class of lipid chemical mediators), J. Biol. Chem., 254:9355 (1979).

23. P.M. Henson, Release of vasoactive amines from rabbit platelets induced by sensitized mononuclear leukocytes and antigen, J. Exp. Med., 131:287 (1970).

24. R.P. Siraganian and A.G. Osler, Destruction of rabbit platelets in the allergic response of sensitized leukocytes. II. Evidence of basophil involvement, J. Immunol., 106:1252 (1971).

25. J. Benveniste, P.M. Henson, and C.G. Cochrane, Leukocyte-dependent histamine release from rabbit platelets. The

role of IgE, basophils and a platelet-activating factor, J. Exp. Med., 136:1356 (1972).

26. R.N. Pinckard, R.S. Farr, and D.J. Hanahan, Physiochemical and functional identity of rabbit platelet-activating factor (PAF) released in vivo during IgE anaphylaxis with PAF released in vitro from IgE sensitized basophils, J. Immunol., 123:1847 (1979).

27. G. Camussi, C. Tetta, M.C. Deregibus, F. Bussolino, G. Segoloni, and A. Vercellone, Platelet-activating factor (PAF) in experimentally induced rabbit acute serum sickness: role of basophil-derived PAF in immune complex deposition, J. Immunol., 128:86 (1982).

28. G. Camussi, C. Tetta, F. Bussolino, F. Caligaris-Cappio, C. Masera, and G. Segoloni, Mediators of immuno complex induced aggregation of polymorphonuclear neutrophils. II. Platelet-activating factor as the effector substance of immune-induced aggregation, Int. Archs. Allergy Appl. Immun., 64:25 (1981).

29. G. Camussi, J.M. Mencia-Huerta, and J. Benveniste, Release of platelet-activating factor and histamine. I. Effect of immunecomplexes, complement and neutrophils on human and rabbit mastocytes and basophils, Immunology, 33:523 (1977).

30. R.L. Lewis, D.R. Morton, S.I. Wasserman, F.H. Valone, R.H. Rubin, and F.K. Austin, The release of four mediators of immediate hypersensitivity from human leukemic basophils, J. Immunol., 114:87 (1975).

31. G. Camussi, M. Aglietta, R. Coda, F. Bussolino, W. Piacibello, and C. Tetta, Release of platelet-activating factor (PAF) and histamine. II. The cellular origin of PAF: monocytes, polymorphonuclear neutrophils and basophils, Immunology, 42:191 (1981).

32. G.Z. Lotner, J.M. Lynch, S.J. Betz, and P.M. Henson, Human neutrophil-derived platelet-activating factor, J. Immunol., 124:676 (1980).

33. R.N. Pinckard, L.M. McManus, and D.J. Hanahan, Chemistry and biology of acetyl glyceryl ether phosphorylcholine (platelet-activating factor), in: "Advances in Inflammation Research," Vol. 4, G. Weissman, ed., Raven Press, New York, p. 147 (1982).

34. J. Benveniste, J.P. LeCouedic, and P. Kamoun, Aggregation of human platelets by platelet-activating factor, Lancet, i:344 (1975).

35. G. Camussi, F. Bussolino, C. Tetta, J. Benveniste, and A.
     Vercelone, Platelet-activating factor and glomerulonephritis,
     in: "Homeostasis, Prostaglandins and Renal Disease," G.
     Remuzzi, et al., eds., Raven Press, New York, p. 133 (1980).
36. L. McManus, D.J. Hanahan, and R.N. Pinckard, Human platelet
     stimulation by acetyl glyceryl ether phosphorylcholine,
     J. Clin. Invest., 67:903 (1981).
37. A.J. Marcus, L.B. Safier, H.L. Ullman, K.T.H. Wang, M.J.
     Broekman, B.B. Weksler, and K.L. Kaplan, Effects of acetyl
     glyceryl ether phosphorylcholine on human platelet function
     in vitro, Blood, 58:1027 (1981).

HEMOSTATIC PARAMETERS AND

ISCHEMIC HEART DISEASE

M. Cortellaro, C. Boschetti,
G. Moreo, and V. Antoniazzi

University of Milan
Medical Clinic I
Via F. Sforza 35
Milan, Italy

INTRODUCTION

Ischemic heart disease (IHD) is caused by a discrepancy in oxygen supply and demand in the myocardium. Interest in platelet function studies in IHD patients has been stimulated for many years by the knowledge that coronary thrombosis often accompanies myocardial infarction (MI), and this has been increased by discoveries regarding both the platelets' role in the pathogenesis of atherosclerosis and the biologic activity of platelet thromboxane A_2 (TXB_2) in the genesis of coronary angina and MI.

The platelet-vessel interaction in patients with coronary sclerosis is considered to increase TXA_2 secretion with consequent vasoconstriction and platelet aggregation (1), as well as secretion from the α-granules of β-thromboglobulin (β-TG) which inhibits PGI_2 synthesis (2), platelet factor 4 (PF_4) with anti-heparin activity (3), and the atherogenetic platelet-derived growth factor (PDGF) (4,5).

Based on the above observations, many laboratory studies have been performed in an attempt to find hemostatic tests which reliably reflect the role played by the hemostatic system in the pathogenesis of IHD, in order to use them to identify patients at risk

for such diseases and to monitor drug effects. In vitro tests have
been used, such as aggregometry, and those on isolated platelets,
which have the defect of not taking into account blood flow,
integrity of the vessel wall, or the coagulation and fibrinolytic
systems. Tests used to measure platelet activity in vivo include
the platelet survival test, measurement of circulating platelet
aggregates (CPA), measurement of specific proteins released by
the platelets such as β-TG and PF_4, and measurement of thromboxane
B_2 (TXB_2). The majority of studies have been performed on peripheral
blood samples from populations of subjects with IHD in quiescent
or symptomatic stages. Recently, hemostatic modifications have
been investigated directly at intracoronary sites.

LONG-TERM HEMOSTATIC FUNCTION CHANGES IN MYOCARDIAL INFARCTION
POPULATION

Expensive and sophisticated clinical trials on secondary MI
prevention will be simplified only when it becomes possible to select
a population at risk for coronary events of probable thromboembolic
nature, and when it is known which functional metabolic stages of
of the platelets must be influenced so that treatment may be
expected to be effective.

As part of the Anturan Reinfarction Italian Study (ARIS) (6),
a controlled prospective study on the efficacy of sulfinpyrazone
(S), we designed a prospective laboratory study to investigate if
a combination of two or more abnormal patterns of hemostatic parameters existed in the placebo group (P), and if some hemostatic
parameters had a predictive value and could provide valid data for
the selection of patients at risk for coronary thrombosis. The
results of this study have been reported elsewhere (7,8).

Tests which must be repeated often and on large populations
must be simple; and we used relatively simple tests to investigate
the trend of some parameters: bleeding time (BT) (9), platelet
count (Thrombocounter, Technicon), platelet retention (PR) (10),
PF_4 release (11), plasmatic heparin neutralizing activity (HNA),
espressed as heparin thrombin clotting time (HTCT) (12), plasma
β-TG (RIA kit, Radiochemical Centre, Amersham), platelet aggregation
(PA) (13), and fibrinogenemia (Fg) (14). These parameters were
studied in a group of 186 patients selected randomly from the ARIS
series of 727 MI patients who, as reported (6), were put on treatment with S (95 cases) (400 mg/2/die) or P (91 cases) 15-25 days

Table I. Hemostatic parameters in patients 15–25 days after MI and in controls (values expressed as mean $\pm$ S.D.)

	Sulfinpyrazone subsample		Placebo subsample		Controls
	A	B	A	B	
Platelet count x 10^3/mm^3	284.82 ±77.71	281.67 ±74.75	305.46 ±102.69	274.89 ±98.59	264.48 ±76.21
Bleeding time (sec)	148•• ±60.78	150.98•• ±55.56	125.63•• ±33.33	148.64•• ±57.16	264.11 ±112.98
Platelet retention (%)	23.71 ±10.94	26.12 ±12.45	26.93 ±18.17	26.24 ±17.13	20.96 ±10.08
Fibrinogen (mg/dl)	269.13•• ±66.85	360.5•• ± 113.24	343.62•• ±101.67	334.13•• ±84.58	237.85 ±35.77
HTCT (sec)	27.52•• ±13.99	30.98• ±14.25	32.52• ±17.07	28.75•• ±14.32	36.31 ±11.31
PF$_4$ (U/ml)	0.39•• ±0.11	0.36•• ±0.13	0.36•• ±0.11	0.39•• ±0.12	0.28 ±0.12
β-TG (ng/ml)	84.35• ±38.41	115.45• ±45.07	91.14• ±39.56	120.7• ±43.77	30.8 ±11.23
ADP 0.2 μM max 1st wave	39.74 ±13.33	44.15• ±12.35	42.12 ±16.12	43.63• ±13.66	37.45 ±15.91
slope 1st wave	66.04 ±30.88	70.68•• ±22.03	66.75 ±33.47	65.83 • ±25.29	55.55 ±23.05
Δ E$_{480}$	35.61 ±28.04	46.80 ±28.91	34.82 ±28.36	50.26 ±29.73	39.64 ±27.59
Collagen 1.25 μg/ml LP	77.79 ±14.19	89.61 ±37.58	88.93 ±24.13	85.64 ±26.49	86.76 ±24.49
slope	61.29•• ±19.07	57.53•• ±23.22	59.07 •• ±19.09	59.16•• ±20.78	44.24 ±23.46
max.	80.21•• ±10.62	76.26•• ±17.91	76.89 • ±12.26	76.65 •• ±16.61	64.84 ±23.58

A ⩽ 50 years   Comparison vs controls ("F" test):   •• p < 0.05
B > 50 years   •• p < 0.01

days after infarct episode. All the tests were performed when the
patients were enrolled (baseline values) and 1, 6, and 12 months
after enrollment.

A more sophisticated battery of tests were performed 6 months
after enrollment in another group of patients also from the ARIS,
aged between 30 and 40 years, as follows: platelet malondialdehyde
(MDA) production (15), platelet production time (PPT) (16), factor
VIIIR:Ag (17), and plasma levels of PF_4 (RIA kit, Abbott Laboratories) and β-TG.

The results confirmed the existence of hyperactive platelets
in the early post-MI phase (within 15-25 days) (Tab. 1), expressed
in vivo by a significantly shortened BT and by significantly increased plasma β-TG, and in vitro by a significant increase in
PF_4 release and HNA, and by platelet sensitivity to threshold
concentrations of ADP (0.2 μM) and collagen (1.25 μg/ml). Platelet
retention was also increased but not significantly.

Only observation of the pattern of the various parameters in
the individual patients at different times after MI could clarify
if the modifications observed represented a phenomenon secondary
to the acute phase of the necrosis, or if they were characteristics
of the patients and thus of the ischemis disease in quiescent phase.
Fg, which was still significantly increased only at month 1, behaved
like an "acute phase reaction" protein. There were no longer differences in the platelet aggregating response to threshold doses
of ADP and collagen between the two treatment subsamples or between
the patients and controls. BT persisted significantly shortened
and PF_4 release increased at all observation times. BT shortening
could reflect the increased number of circulating megathrombocytes
due to an increase in platelet turnover. HTCT, proposed by O'Brien
(18) as an index of heparin-neutralizing platelet activity, could
partially reflect the increased secretion of PF_4, stored in the
α-granules together with β-TG. However, the absence of correlation
between HTCT and PF_4 release at each observation time and between
HTCT and plasma β-TG levels seems to indicate that HTCT cannot be
considered a reliable index of increased release of α-granules or
of platelet hyperactivity in vivo, but rather as an aspecific test
which depends on various plasma coagulation factors. The close
correlation found between this test and Fg concentration (p<0.01)
seems to confirm such an interpretation. On the basis of the data

reported here, it is not possible to establish that increased PF_4 release depends on increased platelet sensitivity to collagen stimulus or on increased intra-platelet PF_4 concentration.

Results of the tests at 6 months in the young patients showed a significant increase in platelet turnover in the P group vs. the control group ($p < 0.01$) and vs. the S group ($p < 0.01$), using both the linear and exponential models, while the S group did not differ from the controls with any of the three models of analysis of the MDA regeneration curves (Fig. 1). The fact that in the determination of PPT the exponential curve provides the best fit for the values of MDA production in the P group suggests that in such patients the removal of platelets from the circulation is the result not only of platelet senescence (in this case, the linear curve should provide the best fit, and the S group presents this characteristic), but also of a consumption mechanism with consequent reduction of platelet survival, as already reported in IHD (19-21). In the P subsample, but not in the S, there was a significant increase of factor VIIIR:Ag vs. the controls ($p < 0.05$), whereas, the plasma levels of β-TG and PF_4 were significantly increased in both the treatment groups (Tab. 2). The high levels of β-TG and PF_4 do not correlate with PPT values found in our patients, as already reported (22). On the other hand, the PPT values correlated inversely with the factor VIIIR:Ag levels ($r = -0.626$, $p < 0.05$). Therefore, since VIIIR:Ag is localized not only in the α-granules like β-TG and PF_4 (23) but also in endothelial cells (24), it may be supposed that increase in VIIIR:Ag is more an expression of its secretion by the damaged endothelial cells. In fact, S was shown to be unable to reduce α-granules secretion and thus β-TG and PF_4 levels, whereas, it did reduce VIIIR:Ag levels possilby due to its hypothesized "protective" action of the endothelium (25). Moreover, since this second group of patients was relatively young (30-40 years), could VIIIR:Ag be considered as a marker of early athero-sclerotic lesions?

It was thought that considering all our above results together, a multifactorial pattern would be suggested of hemostatic parameters (BT, HNA, PF_4 release, β-TG, PF_4, PPT, and VIIR:Ag) to be used to select patients at risk for thromboembolic events characteristic of the natural history of IHD. However, the correlations sought between such parameters and the clinical endpoints presented by the patients studied were disappointing. The subjects in whom clinical endpoints occurred were not characterized by a higher

Table II. Hemostatic Parameters in Young Patients 6 Months after MI and in Controls (Values Expressed as Mean $\pm$ SD)

	Sulfinpyrazone Subsample	Placebo Subsample	Controls
MDA (nmoles x 3 x 10^8 platelets)	3.02 ± 0.13	2.85 ± 0.54	2.4 ± 0.52
β-TG (ng/ml)	68.9 ± 37.71°	98.9 ± 67.7°	35.11 ± 13.93
PF$_4$ (ng/ml)	26.11 ± 23.8••	43.11 ± 39.1°	7.29 ± 4.27
VIIIR:Ag (%)	107.11 ± 34.46	142.13 ± 39.81°	92.29 ± 23.16
VIII:C (%)	86.9 ± 15.1•	111.5 ± 15.38	101.71 ± 33.85
VIIIR:Ag/VIII:C	1.25 ± 0.42	1.28 ± 0.37	0.97 ± 0.29

Comparison ("F" test):
(°) Sulfinpyrazone - placebo vs. controls $p < 0.05$
(•) Sulfinpyrazone vs. placebo $p < 0.01$

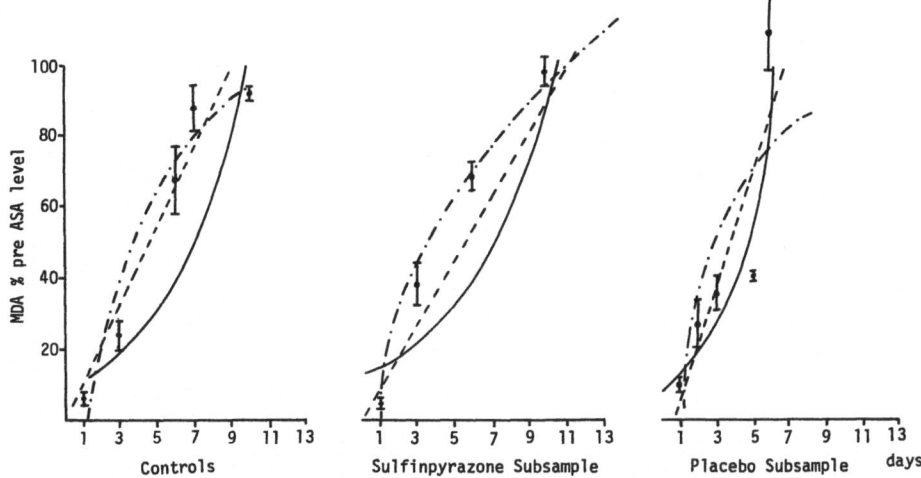

Fig. 1.   Time course of MDA production in MI patient groups
and in controls after aspirin (pre-aspirin value
taken as 100%) according to linear (- - -), exponential (———), and logarithmic(-.-.-.)
models.

incidence of hemostatic abnormalities than the groups of patients
who did not present endpoints.

Nichols et al. (26) have recently studied PF_4 and β–TG as
markers of α-granules release in vivo (27) as well as fibrinopeptide A (FPA), the first peptide released from fibrinogen by
thrombin (28), in 82 patients with angiographically documented
IHD. They found that these proteins were not increased in the
patients who had not presented an infarction. On the other hand,
as in our study, these authors also found significant increase in
PF_4 and β–TG levels in the patients who had had an MI six months
previously, but FPA in the normal range. These levels did not
correlate with the extent of coronary lesions, but they correlated
with the ventricular regional dysfunction: these data support
platelet activity at the level of the damaged ventricular wall,
rather than at the level of the coronary atheromatous lesions.

In the studies in which PF_4 and β–TG in IHD patients were
found to be increased (27,30), the patients with previous MI were
not analyzed separately. Sobel et al. (31) also did not find
increased FPA levels in the peripheral blood of IHD patients,
suggesting that thrombin generation is not responsible for release

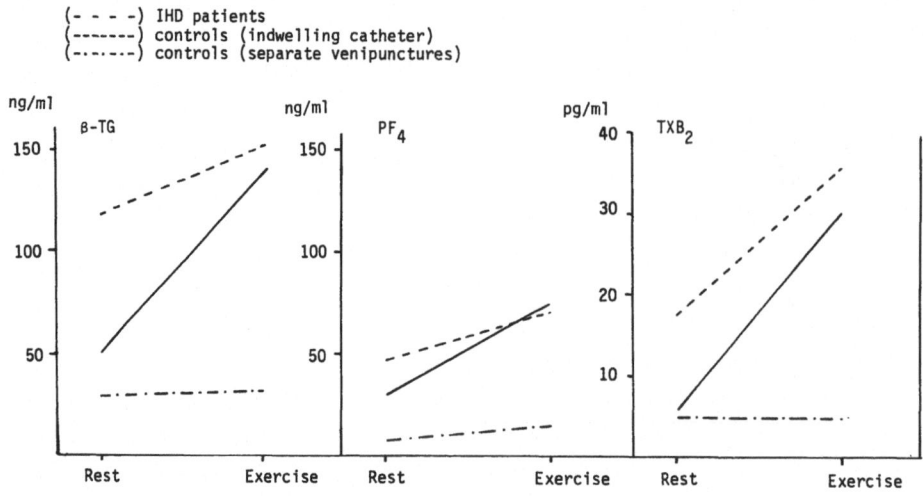

Fig. 2. Plasma β-TG, PF$_4$, and TXB$_2$ levels at rest and after
         exercise test in IHD patients and in controls.

of platelet constituents, since the thrombin concentration needed
for β-TG and PF$_4$ secretion is 100 times more than that for FPA
release.

The FPA findings reported in the peripheral blood could probably
be connected with the existence of a ventricular aneurysm and mural
thrombi. Neri Serneri et al. (32) studied the correlation between
increased FPA, β-TG, and CPA levels and the clinical characteristics
of IHD patients. No correlations were found with the severity of
coronary angiographic lesions or with metabolic risk factors. The
authors suggest that there are other factors responsible for
platelet activation in addition to vessel lesions.

MODIFICATIONS OF HEMOSTATIC PARAMETERS AND EXERCISE TESTING

Enormous interest has been generated by the study in which
Green et al. (33) reported a predictive significance for PF$_4$,
which according to their data correlates with the degree of
exercise testing modifications of the S-T segment. Recently,
Stratton et al. (34) have questioned these conclusions. In IHD
patients undergoing treadmill exercise tests, they obtained
fewer increases of PF$_4$ and β-TG than in normal patients, and they
stressed that the greatest modifications were found using an
indwelling catheter for repeated blood sampling, whereas, the

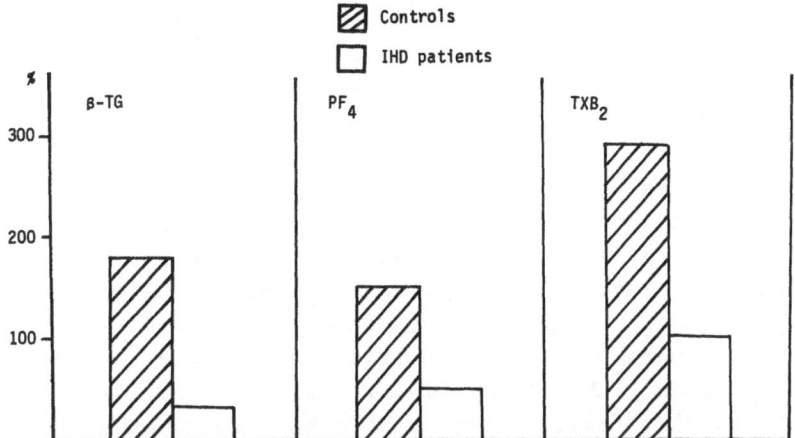

Fig. 3. Percent increase of plasma β-TG, PF_4, TXB_2 after exercise test in IHD patients and in controls.

modifications were insignificant with separate venipunctures during the exercise test.

We studies 12 normal subjects and 8 subjects with at least one coronary artery stenosis documented angiographically. They all underwent symptom-limited maximal treadmill exercise testing. At rest and at maximum effort, the plasma levels of TXB_2, β-TG, and PF_4 were determined. All subjects had an indwelling catheter in the right arm, kept open with saline without heparin, for plasma sampling. The results are shown in Figs. 2 and 3. In the control group, the basal levels of β-TG and PF_4 were higher than the normal values of our laboratory (β-TG 50.73 $\pm$ 26.16 vs. 37.9 $\pm$ 16.5 ng/ml; PF_4 29.7 $\pm$ 28.72 vs. 4.96 $\pm$ 3.21 ng/ml). At the end of the exercise test, the β-TG and PF_4 levels had increased compared with basal values (β-TG 139.45 $\pm$ 68.07 ng/ml; PF_4 76.4 $\pm$ 38.38 ng/ml; mean increase: β-TG 174%, PF_4 157.2%). TXB_2 levels were undetectable basally (<0.05 ng/ml). After testing, measurable levels were found in 10 subjects (mean 300 $\pm$ 180 pg/ml). The subjects with IHD had higher basal levels of β-TG (118.25 $\pm$ 73.46 ng/ml) and PF_4 (47.06 $\pm$ 39.72 ng/ml) than the control group, but the percentage increases after testing were lower than those found in the control group, (β-TG 153.0 $\pm$ 22.91 ng/ml; PF_4 73.75 $\pm$ 36.99 ng/ml; increases in β-TG of 29.4% and in PF_4 of 56.7%). The TXB_2 levels at rest could be detected in 5 out of 8 patients (mean 178 $\pm$ 199 pg/ml), and after testing in 7 out of 8 patients (mean 359 $\pm$ 269 pg/ml). The mean

increase of TXB_2 levels after testing was 101.7%. Particularly
high TXB_2 levels were found in one subject who at ventriculography
presented an aneurysm of the ventricle wall with parietal thrombi:
TXB_2 at rest 650 pg/ml, after testing 700 pg/ml.

The elevated values of β-TG and PF_4 at rest also found in
normal subjects led us to suspect the presence of platelet
activation with release of α-granules due to the method of withdrawing blood. Therefore, the exercise test was done in three other
normal subjects, and blood samples taken at rest and after testing
by separate venipunctures. The resulting values were similar ro the
normal values of our laboratory (Fig. 2), thus supporting Stratton
et al.'s observation (34). These authors also showed that the
increases of PF_4 and β-TG correlated with the plasma levels of
epinephrine in relation to the duration of exercise, thus dis-
qualifying the release of these plasma proteins as a marker of
myocardial ischemia.

INTRACORONARY CHANGES OF TXB_2 IN IHD PATIENTS

A series of recent observations on the mechanisms responsible
for angina and infarction suggest that the clinical manifestations
of IHD are the result of an interaction between functional factors,
capable of interfering transiently with the myocardial flow, and
atherosclerotic coronary obstruction. Our understanding of IHD has
been rapidly increasing since coronary spasm was revalued as the
functional factor, superimposed or not on coronary sclerotic
lesions of varying extent, triggering episodes of spontaneous
angina and myocardial infarction. In addition to reflex autonomous
mechanisms and mechanisms linked with circulating humoral agents,
it has been hypothesized that a primary role is played by platelet
prostaglandin endoperoxides converted into TXA_2. This potent
vasoconstrictor, in addition to promoting platelet aggregation
which would worsen the delay in the coronary microvascular flow,
might induce spasm directly.

Some studies have been performed in recent years to clarify
the relationship between the coronary production of TXA_2 and angina
episodes. The primary role of platelet aggregation was favored by
Mehta et al.'s data (35), which showed a fall in the number of
platelets, as well as the generation of circulating aggregates
during the passage of blood through the damaged coronary bed.

The fall in platelets was not confirmed by Dalal et al. (36) or
Simonsen et al. (37) either in unstable or stable angina. Moreover,
these authors and Robertson et al. (38) reported the formation of
intracoronary platelet aggregates only during attacks of variant
angina. Elevated levels of TXB_2 were found (39–41) also in quiescent
clinical conditions, in the peripheral blood, and coronary sinus
of patients with variant angina. These levels increased considerably
during angina attacks. In patients with classical angina, the
peripheral levels were scarsely detectable, whereas, varying increases were found both in the coronary sinus and the peripheral
blood during atrial pacing-induced ischemia, the maximum after 5–10
minutes. Sobel et al. (42) did not find any relation between slight
increases in TXA_2 in the peripheral blood and episodes of effort
angina, and Robertson et al. (38) found that the increases in TXA_2
in the coronary sinus of subjects with spontaneous angina did not
coincide with the onset of ischemia. More recently, Hirsh et al.
(43) found TXB_2 coronary sinus/aorta ratios increased especially
if the angina pain had occurred 24 hours previously, both in effort
angina and in unstable angina. Furthermore, they were able to
ascertain that the levels in the aorta did not correlate with the
number of stenosed coronary vessels.

We determined the plasma levels of TXB_2 simultaneously at the
coronary sinus (CS), the ascending aorta (Ao), and an antecubital
vein before and after atrial pacing in a group of 21 patients with
classical angina, all with positive exercise test, and in a control
group consisting of 6 patients with atypical chest pain. The 21
angina patients had, at coronary arteriography, at least one
critical stenosis (>75%) in one or more coronary arteries, whereas,
no stenoses were found in the control group. Six of the 21 angina
repeated the study after 15 days' treatment with sulfinpyrazone
(800 mg/die). The results of this investigation, which are in press,
may be summarized as follows: the TXB_2 levels at the CS, Ao, and
peripheral vein were not statistically different between the two
groups of subjects. Atrial pacing caused a significant ($p<0.05$)
increase in TXB_2 levels at the CS, but not at the Ao or peripheral
vein in the angina patients. On the contrary, in the controls,
no differences were shown between levels at the three sites. The
TXB_2 CS/Ao ratio correlated significantly ($r:0.48$, $p<0.05$) with the
coronary score, but not with the number of stenosed coronary vessels.
After 15 days' treatment with sulfinpyrazone, the six angina patients
showed a reduction in the CS and Ao levels of TXB_2 at atrial pacing

peak; however, this reduction was not accompanied by an improvement
of the ischemic threshold. Analysis of variance performed on all
the results in both groups did not show any other significant
differences.

All these studies considered together seem to contrast with the
hypothesis that the stimulus to produce TXA_2 may be generated
by the passage through an atherosclerotic bed, whereas, they seem
to indicate that the ischemic episode, either because of increased
myocardial metabolic demand or coronary spasm, induces alterations
in the circulation which lead to sensitivation of the platelets,
with consequent release of TXA_2.

TXB_2 levels have been suggested (44) as a marker of coronary
failure in trail protocols of antiaggregants. In agreement with
Hirsh et al. (43), we believe that this proposal is not valid,
since increased TXB_2 levels are not found in the peripheral blood
of patients with quiescent classical angina.

From Hirsh et al.'s study (43), furthermore, it seems there is
no role for PGI_2. In fact, no differences were observed between
the CS/Ao ratio for this substance in the groups of patients studied.
Moreover, these authors observed that a deficit of coronary PGI_2,
in the presence of an elevated transcardiac concentration of TXB_2,
did not seem to be responsible for the syndrome of unstable angina.
In addition, Chierchia et al. (45) showed that PGI_2 administration
did not prevent spasm, and these authors (46) and Robertson et al.
(47) showed that the inhibition of TXA_2 production with low-dose
aspirin did not reduce the incidence of angina episodes. These
results seem to weaken the hypothesized significance of the balance
between the circulating factors TXA_2 and PGI_2 in triggering the
ischemic picture. Our study, which showed that sulfinpyrazone
did not improve the atrial pacing ischemic threshold although it is
able to reduce TXB_2 generation, provides further data indicating
that increases of this prostaglandin metabolite are an epiphenomena
secondary to ischemia. To make the problem still more complex, local
hypersensitivity (generated by a change in the TXB_2/PGI_2 ratio
intrinsic to the endothelial wall) of the vessel wall to a variety
of stimuli has recently been postulated on the basis of various
observations. Such abnormalities of the vessel wall may also be
systemic, thus explaining the coexistence of episodes of migraine
and Raynaud's phenomenon in patients with variant angina.

SIGNIFICANCE OF MODIFICATIONS IN HEMOSTATIC PARAMETERS IN IHD

The modifications demonstrated in hemostatic parameters in the quiescent phase of IHD were interpreted until recently as the consequence of morphologic and biochemical modifications occurring in the platelets after repeated passage through coronary beds with damaged walls. Studies have, therefore, been oriented towards searching for correlations between such modifications and the classical risk factors for atherosclerosis. Most of these studies have produced discouraging results, and the results of attempts at finding a correlation between hemostatic parameters and clinical events in IHD have been still more disappointing. On the other hand, these results may be explained by recent evidence of an absence of correlations between peripheral and localized modifications of some hemostatic parameters and the extent of the coronary angiographic lesion. Numerous observations which emphasize the importance of functional factors in the pathogenesis of angina and infarction now indicate that the prognosis of critical events of IHD depends on the severity of these abrupt factors as well as the degree of impairment of the coronary blood flow caused by the chronic coronary lesion.

Modifications of hemostatic parameters, if they are a characteristic of an IHD population, do not appear to be determining or indispensable factors for the occurrence of the clinical events in IHD. Available laboratory tests, more or less sophisticated, either on peripheral blood and/or on samples of intracoronary blood, are useful and appropriate to show the existence of platelet hyperactivity and of a state of "hypercoagulability" in IHD, but they still do not have a predictive role of critical events in either angina or infarction, and least of all in sudden death. They may help to define a population (not a single individual) at risk of coronary thromboembolic events, but such events may occur because of the sudden appearance of a mysterious functional trigger involving both normal and hyperactive platelets and normal or damaged vessels.

In other words, IHD, with or without angina, with or without infarction, may be two different clinical entities. Obviously, since PF_4 neutralizes heparin's anticoagulant effect, β-TG possibly inhibits PGI_2 production, and TXA_2 is a potent vasoconstrictor and aggregates the platelets, the chronic release of such substances cannot not favor sclerosis and thrombosis in the coronary vessels. If patients after myocardial infarction and ischemia present a prolonged increase of these proteins accompanied or not by reduced platelet survival, it can reasonably be suggested that such patients constitute a group at risk which could benefit from preventive treatment with antiaggregants.

REFERENCES

1. M. Hamberg, J. Svensson, and B. Samuelsson, Thromboxanes: a new group of biologically active compounds derived from prostaglandin endoperoxides, Proc. Natl. Acad. Sci. USA, 72:2294 (1975).
2. W. Hope, T.J. Martin, C.N. Chesterman, and F.J. Morgan, Human β-thromboglobulin inhibits PGI_2 production and binds to a specific site in bovine aortic endothelial cells, Nature, 282:210 (1978).
3. R.I. Handin and H.J. Cohen, Purification and binding properties of human platelet factor 4, J. Biol. Chem., 251:4273 (1975).
4. L.D. Witte, K.L. Kaplan, H.L. Nossel, G.A. Lager, H.J. Weiss, and D.S. Goodman, Studies of the release from human platelets of the growth factor for cultured human arterial smooth muscle cells, Circ. Res., 42:402 (1978).
5. R. Ross, J. Glomset, B. Kariya, and L. Harker, A platelet-dependent serum factor that stimulates the proliferation of arterial smooth muscle cells in vivo, Proc. Natl. Acad. Sci. USA, 71:1207 (1974).
6. Anturan Reinfarction Italian Study, Sulfinpyrazone in post-myocardial infarction, Lancet, 1:237 (1982).
7. M. Cortellaro, C. Boschetti, G. Fassio, L. Baroni, and E.E. Polli, Haemostatic function changes in a trial on the secondary prevention of myocardial infarction with sulfinpyrazone, Acta Haematol., 65:193 (1981).
8. M. Cortellaro, C. Boschetti, P. Beggi, and E.E. Polli, In vivo platelet hyperactivity and factor VIII related antigen increase long after myocardial infarction, Scand. J. Haematol., 26:106 (1981).
9. C. Praga, M. Cortellaro, and E. Pogliani, Standardized bleeding time in the drugs interfering with platelet function, in: "Platelet Function in Thrombosis: A Review of Methods," Plenum Press, New York, pp. 149-153 (1972).
10. A.J. Hellem, Platelet adhesiveness in von Willebrand's disease.

A study with a new modification of the glass bead filter method, Scand. J. Haematol., 7:374 (1970).

11. K. Harada and M.B. Zucker, Simultaneous development of platelet factor 4 activity and release of ^{14}C-serotonin, Thromb. Diath. Haemorrh., 25:41 (1971).

12. J.R. O'Brien, J.B. Heywood, and J.A. Heady, A quantitation of platelet aggregation induced by four compounds: a study in relation to myocardial infarction, Thromb. Diath. Haemorrh., 16:752 (1966).

13. J.V.R. Born and M.J. Cross, The aggregation of blood platelets, J. Physiol., 168:178 (1973).

14. M.B. Donati, G. De Gaetano, J. Vermylen, and M. Verstraete, Il Fibrin Polymeration Time (FTP) test: un metodo rapido, semplice e sensibile per la determinazione della fibrino-genemia, Farmaco, 27:214 (1972).

15. S. Villa, M. Livio, and G. De Gaetano, The inhibitory effect of aspirin on platelet and vascular prostaglandins in rats cannot be completely dissociated, Br. J. Haematol., 42:425 (1979).

16. M.J. Stuart, S. Murphy, and F.A. Oski, A simple non-radioisotope technic for the determination of platelet life-span, N. Engl. J. Med., 292:1310 (1975).

17. C. Laurell, Quantitative estimation of proteins by electro-phoresis in agarose gel containing antibodies, Anal. Biochem., 15:45 (1966).

18. J.R. O'Brien, M.D. Etherington, S. Jamieson, P. Lawford, J. Sussex, and S.B. Lincoln, Heparin neutralizing activity test in the diagnosis of myocardial infarction, J. Clin. Pathol., 28:975 (1975).

19. P.P. Steele, H.S. Weiss, H. Davies, and E. Genton, Platelet function studies in coronary artery disease, Circulation, 48:1194 (1973).

20. J.A. Ritchie and L.A. Harker, Platelet and fibrinogen survival in coronary atherosclerosis. Response to medical and surgical therapy, Am. J. Cardiol., 39:595 (1977).

21. M. Cortellaro, C. Boschetti, G. Fassio, and E.E. Polli, A controlled study of the effect of sulfinpyrazone on platelet survival and on platelet bound ^{14}C-serotonin release in patients with previous myocardial infarction, Acta Haematol., 61:68 (1979).

22. P. Han, A.G.G. Turpie, E. Genton, J. Hirsh, and M. Gent, Negative correlation between platelet survival, β-thrombo-globulin and platelet aggregate ratio in patients with

coronary artery disease, Circulation, 58(Suppl. 11):116
        Abstr. (1978).

23. J.C. Giddings, L.R. Brookes, F. Piovella, and A.L. Bloom,
    Immunohistological comparison of platelet factor 4 (PF_4),
    fibronectin (Fn), and factor VIII-related antigen (VIIIR:Ag)
    in human platelet granules, Br. J. Haematol., 52:79 (1982).

24. J.H. Rand, R.E. Gordon, I.I. Sussman, S.V. Chu, and V. Solomon,
    Electron microscopic localization of factor VIII-related
    antigen in adult human blood vessels, Blood, 60:627 (1982).

25. L.A. Harker, R.J. Wall, and J.M. Harlen, Sulfinpyrazone
    prevention of homocystein-induced endothelial cell injury
    and arteriosclerosis, Clin. Res., 26:554 (1978).

26. A.B. Nichols, J. Owen, K.L. Kaplan, R.R. Sciacca, P.J. Cannon,
    and H.L. Nossel, Fibrinopeptide A, platelet factor 4, and
    β-thromboglobulin levels in coronary heart disease, Blood,
    60:650 (1982).

27. K.L. Kaplan, H.L. Nossel, M. Drillings, and G. Lasznik, Radio-
    immunoassay of platelet factor 4 and β-thromboglobulin:
    development and application to studies of platelet release
    reaction to fibrinopeptide A generation, Br. J. Haematol.,
    39:129 (1978).

28. H.L. Nossel, M. Ti, K.L. Kaplan, K. Spanondis, T. Soland, and
    Y.P. Butler, The generation of fibrinopeptide A in clinical
    blood samples. Evidence for thrombin activity, J. Clin.
    Invest., 58:1136 (1976).

29. S.P. Levine, J.A. Lindenfeld, J.B. Ellis, N.M. Raymond, and
    L.S. Krentz, Increased plasma concentrations of platelet
    factor 4 in coronary artery disease, Circulation, 64:626
    (1981).

30. G.C. White and A.A. Marony, Platelet factor 4 levels in coronary
    artery disease, J. Lab. Clin. Med., 97:369 (1981).

31. M. Sobel, E.W. Salzman, G.C. Davies, R.I. Handin, J. Sweeney,
    J. Ploetz, and G. Kurland, Circulating platelet products
    in unstable angina pectoris, Circulation, 63:300 (1981).

32. G.G. Neri Serneri, G.F. Gensini, R. Abbate, C. Mugnaini, G.
    Favilla, C. Brunelli, S. Chierchia, and O. Parodi,
    Increased fibrinopeptide A formation and thromboxane A_2
    production in patients with ischemic heart disease:
    relationships to coronary pathoanatomy, risk factors, and
    clinical manifestations, Am. Heart J., 101:185 (1981).

33. L.H. Green, E. Seroppian, and R.I. Handin, Platelet activation
    during exercise-induced myocardial ischemia, N. Engl. J.
    Med., 302:193 (1980).

34. J.R. Stratton, T.W. Malpass, J.L. Ritchie, M.A. Pfeifer, and L.A. Harker, Studies of platelet factor 4 and beta-thromboglobulin release during exercise: lack of relationship to myocardial ischemia, Circulation, 66:33 (1982).

35. J. Mehta, P. Mehta, C.J. Pepine, and C.R. Conti, Platelet function studies in coronary artery disease. VII: Effect of aspirin and tachycardia stress on aortic and coronary venous blood, Am. J. Cardiol., 45:945 (1980).

36. J.T. Dalal, D.J. Sheridan, A.L. Bloom, and A.H. Henderson, Platelet aggregates and coronary spasm, Lancet, ii:1146 (1980).

37. S. Simonsen, J. Dali, T. Hovig, J. Kjeskus, N.V. Solmon, and E. Thanlow, Myocardial ischemia and platelet reactivity, in: VIII European Congress of Cardiology, Abstr. 126, Paris (1980).

38. R.M. Robertson, D. Robertson, L.J. Roberts, R.L. Maas, G.A. Fitzgerald, G.C. Friesinger, and J.A. Oates, Thromboxane A_2 in vasotonic angina pectoris, N. Engl. J. Med., 304: 998 (1981).

39. R.I. Lewy, L. Wiener, P. Walinsky, A.M. Lefer, M.J. Silver, and J.B. Smith, Thromboxane A_2 in Prinzmetal's variant angina and classical angina pectoris, Clin. Cardiol., 2: 404 (1979).

40. R.I. Lewy, L. Wiener, P. Walinsky, A.L. Lefer, M.J. Silver, and J.B. Smith, Thromboxane release during pacing-induced angina pectoris: possible vasoconstrictor influence on the coronary vasculature, Circulation, 61:1165 (1980).

41. M. Tada, T. Kuruya, M. Imane, K. Kadama, M. Mishima, M. Yamada, M. Inui, and H. Abe, Elevation of thromboxane B_2 levels in patients with classic and variant angina pectoris, Circulation, 64:1107 (1981).

42. M. Sobel, E.W. Salzman, G.C. Davies, R.I. Handin, J. Sweeney, J. Ploetz, and G. Kurland, Circulating platelet products in unstable angina pectoris, Circulation, 63:300 (1981).

43. P.D. Hirsh, L.D. Hillis, W.B. Campbell, B.G. Firth, and J.T. Willerson, Release of prostaglandins and thromboxane into the coronary circulation in patients with ischemic heart disease, N. Engl. J. Med., 304:685 (1981).

44. R.I. Lewy, Thromboxane in ischemic heart disease, N. Engl. J. Med., 305:186 (1981).

45. S. Chierchia, C. Patrono, F. Crea, G. Ciabattoni, R. De Caterina, G.A. Cinotti, A. Distante, and A. Maseri,

Effects of intravenous prostacyclin in variant angina, Circulation, 65:470 (1982).

46. S. Chierchia, R. De Caterina, F. Crea, C. Patrono, and A. Maseri, Failure of thromboxane A_2 blockade to prevent attacks of vasospastic angina, Circulation, 66:702 (1982).

47. R.M. Robertson, D. Robertson, L.J. Roberts, R.L. Mass, G.A. Fitzgerald, G.C. Friesinger, and J.A. Oates, Thromboxane A_2 in vasotonic angina pectoris: evidence from direct measurements and inhibitory trials, N. Engl. J. Med., 304:988 (1981).

CRITICAL REVIEW OF MYOCARDIAL INFARCTION

PREVENTION WITH ANTIAGGREGANTS

E.E. Polli and M. Cortellaro

University of Milan
Clinica Medica I
Via F. Sforza 35
Milan, Italy

Myocardial infarction (MI) is the result of a low process of narrowing of the coronary vessels and an acute thrombotic event. The most rational approach to MI prevention should, therefore, be the prevention of coronary atheroma. However, although experimental data exist which indicate that antiaggregants are effective in preventing the formation of atherosclerotic lesions, and although a clinical study with a 5-year follow-up is presently being performed on the use of antiaggregants in inhibiting the angiographic progression of coronary sclerosis (1), at the present time the prevention of coronary atheroma does not seem to be clinically realistic. For this reason, research has been directed towards prevention of thromboocclusive events in the coronary vessels.

The antiaggregant drugs used until now in the prevention of complications of coronary and cerebral vascular disease modulate some stages in platelet metabolism leading to the formation of mediators (cyclic CMP, arachidonic acid metabolites, calcium ions) of the various platelet reactions. The effects on the synthesis of the various products of the endothelial prostaglandin chain are still not entirely clear.

Aspirin, one of the three antiaggregant drugs which have been most widely studied clinically, acetylates irreversibly the cyclooxygenases. Since it has been hypothesized that platelet cyclo-

oxygenase is more sensitive to aspirin than endothelial cyclooxygenase (2), studies are being performed to try and establish
the minimum dose of aspirin which will inhibit platelet cyclooxygenase, and thus the synthesis of TXA_2 with its aggregant and vaso-
constricting effect, without interfering with endothelial cyclooxygenase and the synthesis of PGI_2 with its antiaggregant and
vasodilating effects. The clinical antithrombotic effectiveness of
low doses of aspirin obviously still has to be demonstrated.

Furthermore, since the consequences of such acetylation persist
in the platelets for all their survival in circulation because
they are unable to synthesize new enzyme (3), whereas, the endothelial cells are able to resynthesize the enzyme rapidly after
withdrawal of the drug, it has been suggested that the aspirin dose
should not be given at longer intervals (every 48-72 hours).
However, enthusiasm for this new therapeutic strategy has been
dampened by recent studies which cast doubt on the importance of the
platelet thromboxane/endothelial prostacyclin balance in thrombogenesis (4,5).

Sulfinpyrazone also acts by inhibiting cyclooxygenase, but this
effect is reversible (6). The effects of platelet function seem to
be mainly connected with the formation of metabolites of the drug
(7). In addition, neither sulfinpyrazone nor its metabolites affect
endothelial cyclooxygenase at doses inhibiting the platelet enzyme
(8,9). This property could explain why sulfinpyrazone, but not
aspirin, normalizes reduced platelet responsiveness (10,11).

The third widely used antiaggregant, the phosphodiesterase
inhibitor dipyridamole, keeps at a high level intraplatelet cyclic
AMP, a nucleotide which inhibits platelet function. It has been
suggested that this drug acts by potentiating the activity of
endogenous PGI_2 (12); if this is true, it makes it difficult to
explain the usefulness of combining aspiring at high doses (1 g/day)
with dipyridamole, in view of present knowledge of aspirin's inhibition of endothelial cyclooxygenase.

What are the expressions of platelet function which must be
influenced by drug treatment for it to be considered effective?

It is still impossible to reproduce entirely the phenomenon of
thrombosis, and not having predictive tests means that in vitro

we can only study partial and isolated aspects of the complex intravital event. Some drugs may inhibit specific stages of platelet aggregation but not change adhesiveness or bleeding time, whereas, others may be only slightly effective in vitro, but are able to correct reduced platelet survival in certain pathological conditions. This means that an in vivo evaluation of its effectiveness is indispensable for a drug to be defined as antithrombotic.

What are the relationships between platelets and the major clinical manifestations of ischemic coronary disease?

At present, the role of platelets in the various forms of angina is the object of very interesting investigations, especially since the discovery that the connection between coronary sclerosis and ischemic events is not so direct. On the basis of a series of observations on the mechanisms responsible for spontaneous angina, it has been hypothesized that this clinical manifestation could be caused by functional factors, such as coronary spasm or platelet aggregation, capable of interfering transiently with the blood flow in the myocardial region, which may or may not have been superimposed on a preexisting coronary atherosclerotic stenosis (13).

Platelets are responsible for the release of factors like TXA_2 which may modulate the coronary spastic response. A local deficit of PGI_2 could enhance the spastic and aggregation response. This would set up a vicious cycle with spasm and aggregation each potentiating the other.

The pathogenesis of sudden death is still an enigma. A thrombus has been observed in about 20% of the cases, whereas, a severe atherosclerotic stenosis has frequently been found. Furthermore, interesting data have been reported by Cobb et al. (14) on patients resuscitated after ventricular fibrillation. Only 19% of the survivors subsequently had an MI. In other words, the majority of the cases had ventricular fibrillation (which may determine sudden death) without a classical MI.

The role that platelets play is still an open question, considering that intraplatelet vasoconstrictive substances (like TXA_2) may determine spasm and fibrillation in a coronary bed which is already stenosed, and that platelet aggregates may be reversible and not shown by histopathologic investigation.

It is generally agreed today that coronary thrombus has a
casual role in myocardial infarction, particularly transmural MI.
This has been demonstrated by labelled fibrinogen studies, by
findings from coronary arteriographies done in the very early post-
infarct stage, and by autopsy data (15,16).

Coronary spasm, which may be superimposed on a subtotal occlusion and cause an infarction, may itself be triggered by
substances resulting from platelet metabolism like TXA_2.

Before turning to the results of the so-called trials of
secondary prevention of MI with platelet antiaggregants which
sprang up in the 70s', there is one general consideration which
deserves mention: these trial were designed "to evaluate the
antithrombotic effectiveness" of the drugs in question. If platelets
play an essential role in the genesis of thrombosis, and if thrombosis is the precipitating factor of myocardial infarction, it is
reasonable to expect antiaggregants to reduce the incidence of
reinfarction as well as of other major thromboembolic events, but
not to affect cardiac death in general or sudden death in particular.
In the various types of cardiac death after MI, the antiaggregants
can be effective only in the cases in which these events constitute
a real thrombotic endpoint. This means that the studies should
have been designed so that case series were selected with low
mortality risk, so as to evaluate in particular the incidence of
major events of a probable thromboembolic nature. All the trials
were presented as prevention of "reinfarction," whereas, in reality
they were designed, except the Italian one, as trials of the
prevention of "mortality," as if reinfarction always means death,
and as if all post-infarction deaths are reinfarctions. The lack
of strict adherence to the primary rationale, shown particularly
by the choice of endpoints, constitutes in our opinion one of the
main reasons for apparent dissimilarities in the results reported
by the various trials. Another methodological aspect may be
responsible for misleading, and therefore criticizable, conclusions -
the "weighing" of the patients included in the study.

A study designed and analyzed according to the criteria of a
"drug efficacy trial" tries to create an ideal population for the
best and most specific evaluation of a drug's activity, therefore,
in order to study a population which can be shown to be sensitive
to the drug effect, non-eligible and non-compliant patients are

Table I. Main Differences ART/ARIS

	ART	ARIS
Enrollment (days)	25-35	15-25
Age (years)	45-70	71
Dosage	4 x 200 mg	2 x 400 mg
Endpoints	Cardiac death	Cardiac death
		Non-fatal MI
		Other thromboembolic
First 7 days	Excluded	Included
Non-compliant	Excluded	Included
Ineligible	Excluded	Included

excluded from the final analysis as well as medical and non-medical
drop outs. The other type of trial design, the "intent-to-treat
trial," attributes to the drug all the events which occur in all
the enrolled patients during the entire observation period of the
stay from the first administration of the drug. This second type
of trial simulates conditions in clinical practice better, but is
is less sensitive in evaluating the specific efficacy of the drug.

An example of how a choice of different endpoints together
with a different "weight" given to the patients enrolled may lead
to apparently different results is provided by the comparison
between the two multicenter studies on sulfinpyrazone performed
in North America, ART (17), and in Italy, ARIS (18) (Table I).
The American authors decided to design a mortality study (not
enrolling patients under 45 years, and not withdrawing patients
in whom non-fatal events occurred, such as reinfarction and stroke).
They applied strict criteria of a drug efficacy trial, excluding
ineligible and non-compliant patients from the analysis, and in
addition, they applied a particular pharmacological ruling
criticizable for many reasons, which consisted in excluding events
which occurred during the first seven days of drug treatment on
the basis of preliminary data indicating that the drug becomes
clinically effective only after seven days (non-analyzable patients).
The Italian authors, for the reasons mentioned earlier, considered

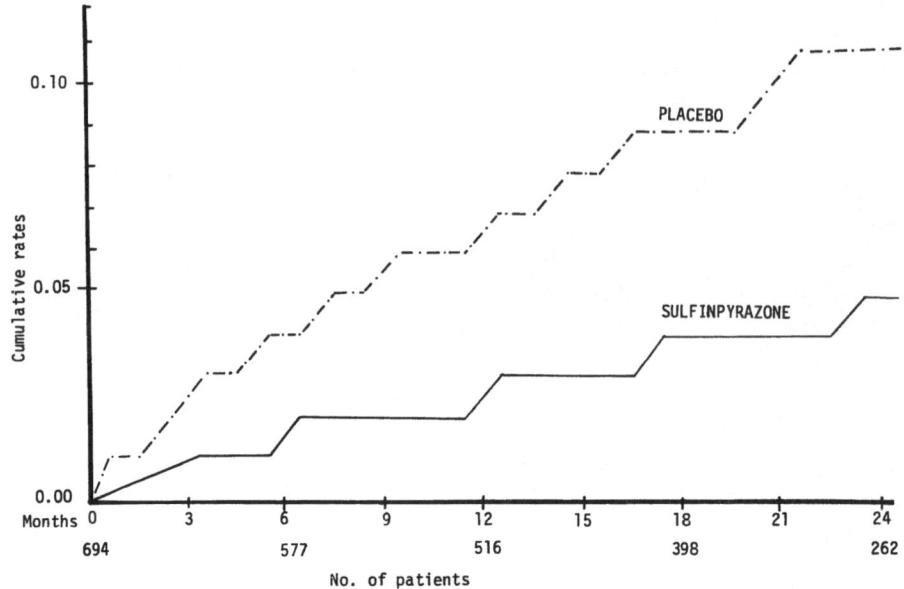

Fig. 1.   Cumulated rates of total reinfarctions

that the major interest of such a trial should not be mortality,
but already in 1975 they chose as endpoints, in addition to the
inevitable cardiac death, non-fatal reinfarctions and other major
thromboembolic events (subjects in whom such an event occurred were
withdrawn from the controlled study).   They also decided to enroll
patients under 45 years.   Moreover, several decisions were taken,
bringing ARIS' design closer to that of an intent-to-treat trial:
all enrolled patients, including non-compliant ones, and all events
from randomization of the patients were to be included in the
analysis of the results.

The ART results were unexpected.   At the end of the trial, there
were 105 cardiac deaths which were considered analyzable, 62 in the
placebo group, and 43 in the sulfinpyrazone group, a 30.6% reduction.
The difference was almost entirely due to the significant reduction
of sudden death (22 in the sulfinpyrazone group vs. 37 in the
placebo group, a 40.5% reduction).   Since half the deaths occurred
in the first 6 month observation period, the beneficial effect of
sulfinpyrazone appears evident in that period (10.3% cardiac deaths
in the placebo group compared with 5.0% in the sulfinpyrazone group,
a 51% reduction), whereas, this difference is not maintained in the
subsequent 18 months (4% in both groups).

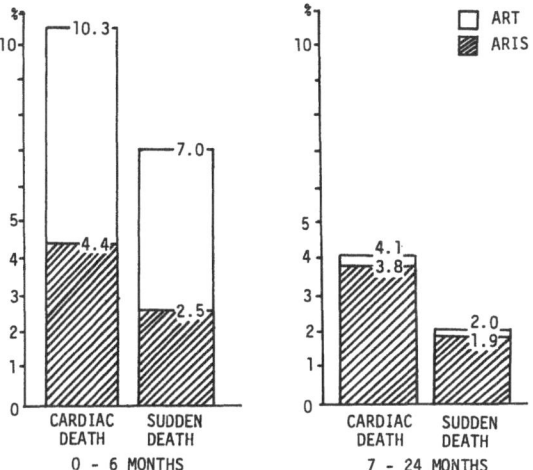

Fig. 2.   Incidence of cardiac death
          and sudden death in 0-6 and
          7-24 month periods in ART
          and ARIS.

In the ARIS, in which the most impressive results were the
56% reduction of fatal plus non-fatal reinfarctions (Fig. 1) and
the 66% reduction of all thromboembolic events, there was no difference in mortality between the two treatment groups.

The apparently different results of the ART and ARIS trials
can be explained by the differences in their respective designs.
The incidence of cardiac deaths in the first 6 months of the ART,
when the drug effect was seen, was greater than in the second
period when no drug effect was evident, and the mortality in the
two trials was in fact similar (Fig. 2). The higher incidence of
the first 6 months of the ART could be explained by the fact that
the risk of death was higher in the population enrolled in the ART
than in that enrolled in the ARIS (Table II); in addition, the fact
that the patients in whom critical non-fatal events occurred were
not withdrawn from the study increased the mortality risk of the
study population. All this is the consequence of the different
endpoints and of differences in the population enrolled. Moreover,
the different "weight" given to the events in the two trials may
have helped to accentuate the differences in their final results.

When the ART results are analyzed according to the FDA critique
(19) (Table III) (considering the 14 deaths which occurred in the

Table II. Main Differences ART/ARIS

	ART	ARIS
Women	213 (13.7%)	60 (8.6%)
Mean age (years)	56.6	53.8
Patients aged 44 years	-	97 (13.9%)
β-blockers	36%	19.4%
Previous MI	323 (20.7%)	85 (12.2%)
Killip class II	34.9%	14.2%
LDH <n	3.32	5.50
n - 2n	19.39	30.01
>2n	77.79	64.49
CPK <n	1.63	4.90
n - 2n	2.58	10.26
>2n	95.79	84.84
SGOT <n	0.75	14.35
n - 2n	4.00	26.59
>2n	95.25	59.06

in the 71 non-eligible patients and the 43 deaths in the non-analyzable patients the reduction in cardiac mortality was 21% (not 30.6%) if all patients and all events are considered, and it was 24% when all the events in the eligible patients are only considered; considering the patients and events clearly defined in the ART protocol, there was a 20% reduction in the entire observation period and 38% (not 48%) in the first 6 months. This is thus the most conservative significant reduction in the American study in the first 6 months of observation when the incidence of mortality was particularly high.

As well as the two trials with sulfinpyrazone, seven other trials have been completed, all with aspirin; one of them also evaluated the aspirin-dypyridamole combination (Table IV).

The first consideration on such trials concerns the lack of homogeneity in the interval between infarction and enrollment (ranging from days to years), resulting in large differences between the various trials in the incidence of events, particularly death which is concentrated in the first months after infarction.

Table III. ART Results According to FDA Critique

Cardiac Mortality	Sulfinpyrazone	Placebo
All patients/all events	65	82
Reduction	21%	NS
Eligible patients/all events	59	78
Reduction	24%	NS
Patients and events clearly defined in ART protocol	55	69
Reduction	20%	p=0.2
First 6 months	27	43
Reduction	38%	p≅0.05

Obviously, although this is not important when the endpoint is the prevention of reinfarctions which are known to be distributed evenly throughout the natural history of post-MI patients, it is important when the endpoint is the prevention of cardiac death. However, all these trials selected cardiac death as the main endpoint. In addition, the daily dose of aspirin ranged from 300 mg to 1.5 g. Despite the enrollment of a total of 11,500 patients in these studies, conclusive results were not obtained.

It has been hypothesized that considering all the trials together, it might be possible to identify a significant trend; however, I do not agree with this type of assessment, since it could mean that trials with different designs and different doses of aspiring might be evaluated in the same way.

In 1974 Elwood et al. (20) studied low daily doses of aspirin in men with recent MI. There was a 25% reduction in mortality, and this difference was greater in a small group of patients enrolled within 6 weeks of their infarct. Since the majority of these latter patients enrolled early were studied in the Cardiff Center, the investigators could not exclude the possibility of a geographic effect.

Table IV. Post-myocardial infarction treatment in randomized trials

Trial	Drug	Dose (mg/die)	Enrollment (after MI)	% reduction (treated vs controls) Mortality	Sudden death	Non-fatal MI
Elwood I (1974)	Aspirin Placebo	300	70 days	24	—	—
CDPA (1976)	Aspirin Placebo	972	85 months	30	19	5
Elwood II (1979)	Aspirin Placebo	900	8 days	17	—	35°
AMIS (1980)	Aspirin Placebo	1000	25 months	+11	+35	22°
PARIS (1980)	Aspirin Aspirin + dipyridamole Placebo	972 972+225	20.2 months	18 16	+27 16	29 19
German Austrian (1980)	Aspirin Phenprocoumon Placebo	1500	35 days	17	34	27
EPSIM (1982)	Aspirin Oral anticoagulant No placebo	Deaths: aspirin group, 72; anticoagulant group, 65; Reinfarctions: " 33; " 20				

° p > 2 standard errors of the difference

The Coronary Drug Project research group (21) enrolled male subjects from two previous studied which had been terminated prematurely. Aspirin had a favorable effect on total cardiac death and coronary incidence. The investigators noted that modifications of cardiac mortality suggested that aspirin had a beneficial effect in secondary prevention; however, the number of patients studied and the follow-up of the individual patients seemed insufficient for definitive conclusions to be drawn. It should be noted that the majority of patients were admitted to the study many years after MI.

In 1979 Elwood and Sweetman (22) reported a second trial in which 50% of the patients were enrolled within 7 days of an MI, the aspirin dose was increased to 900 mg/day, and women were included. The results were considered inconclusive as regards to mortality.

The German-Austrian study (23) randomized subjects soon after MI into aspirin, placebo, and oral anticoagulant treatment groups. The aspirin/placebo randomization was done double-blind. There was again a favorable trend for aspirin for the majority of end-points, but this did not reach any significance. More serious side effects were observed in patients on anticoagulants than in those on aspirin or placebo.

AMIS (24) had the largest number of patients (4,324) and cost over 24 million dollars. Unlike the other trials, no trend was found in total or coronary mortality in favor of aspirin (26 more deaths in the aspirin group). This study showed clearly that 1 g. aspirin daily started 6 months after MI is not effective, as underlined by the authors. Randomization was not satisfactory: the various parameters influencing prognosis were not balanced in the two groups, the aspirin group being penalized.

PARIS (25) was a cooperative study carried out by 16 centers in the U.S.A. and 4 in the U.K. Comparison was between a group on aspirin, one on dipyridamole plus aspirin, and a placebo group of half the size. A dipyridamole group was not included in the study, since it was considered that a dose of 225 mg is ineffective and one of 400 mg badly tolerated. Deaths from all causes, coronary mortality and coronary incidence (coronary deaths plus non-fatal reinfarction) were compared. A favorable effect was shown for the two treatment groups versus placebo (total mortality: aspirin +

dipyridamole 10.7%, aspirin 10.5%, placebo 12.8%), but the differences are not significant. The patients were divided into two
subgroups: those who were admitted within 6 months of their infarct
(20% of the study population), and those who were admitted later.
The results appeared more favorable compared with placebo in the
first 6 months. However, since the subgroups were too small to
allow definitive conclusions, a second study has been started
(PARIS II) with earlier enrollment (patients with their last MI
1-4 months previously).

EPSIM (26) compared a group of 651 patients treated with aspirin
with a group of 652 patients treated with anticoagulants. There
were 65 deaths in the anticoagulant group and 72 in the aspirin
group, whereas, the number of patients with reinfarction was higher
in the aspirin group (33 versus 20). The differences were not
significant. Eighty-two patients on anticoagulants and 139 on
aspirin were withdrawn from the study during follow-up. The antithrombotic effectiveness of the two treatments compared with a
placebo group could not be evaluated with this study.

The reduction of cardiac mortality in all these trials with
antiaggregants reached a maximum of 30% (if the impressive results
obtained for sudden death obtained in the ART are excluded, which
remain to be clarified). However, it must be underlined that in
any case these trials have shown a favorable trend for non-fatal
reinfarction, which reached significance in Elwood II, AMIS, and
ARIS in particular. Such results are very important in light of
the primary rationale of these studies.

What are the pharmacological alternatives to the classical
antiaggregants?

With beta-blockers a significant reduction in post-infarction
mortality has been obtained as well as a certain reduction in non-
fatal reinfarctions (27). The results reported on a possible
antiaggregant effect of such drugs are few and contradictory: there
are even reports of an increased platelet aggregating response to
some agents such as adrenaline (28,29).

A greater antithrombotic effectiveness may be hypothesized
for an antiaggregant/beta-blockers combination, with a greater
reduction of reinfarction. Such action could be synergic in the

reduction of cardiac death, considering the possible antiarrhythmic action of some antiaggregants, and the role of the circulating microaggregates in the pathogenesis of sudden death. A pharmacological study performed by our group in collaboration with the Mario Negri Institute, Milan, on the metaprolol-sulfinpyrazone combination is now in press.

REFERENCES

1. M. Verstraete, Registry of prospective clinical trials, fifth report, Thromb. Haemost., 48:1-5 (1982).
2. N.L. Baezinger, M.J. Dillender, and P.W. Majerus, Cultured human skin fibroblasts and arterial cells produce a labile platelet-inhibitory prostaglandin, Biochem. Biophys. Res. Commun., 78:294-301 (1977).
3. G.J. Roth, N. Stanford, and P.W. Majerus, Acetylation of prostaglandin synthase by aspirin, Proc. Natl. Acad. Sci. USA, 72:3073-3076 (1975).
4. R.M. Robertson, D. Robertson, and L.J. Roberts, Thromboxane A$_2$ in vasotonic angina pectoris: evidence from direct measurements and inhibitory trials, N. Engl. J. Med., 304:998 (1981).
5. S. Chierchia, R. De Caterina, F. Crea, C. Patrono, and A. Maseri, Failure of thromboxane A$_2$ blockade to prevent attacks of vasospastic angina, Circulation, 66:702-705 (1982).
6. M. Ali and J.W.D. McDonald, Effects of sulfinpyrazone on platelet prostaglandin synthesis and platelet release reaction, J. Lab. Clin. Med., 89:868-875 (1977).
7. E.D. Maguire, G.F. Pay, R.B. Wallis, and A.M. White, Prolonged inhibition of ex vivo sodium arachidonate-induced platelet aggregation and malondialdehyde (MDA) production by sulfinpyrazone (Anturane) in man, Thromb. Res., 21:321-327 (1981).
8. J.L. Gordon and J.D. Pearson, Effects of sulfinpyrazone and aspirin on prostaglandin I$_2$ synthesis by endothelial cells, Br. J. Haematol., 64:481-483 (1978).
9. M. Livio, S. Villa, and G. De Gaetano, Long-lasting inhibition of platelet prostaglandin but normal vascular prostacyclin generation following sulfinpyrazone administration to rats, J. Pharm. Pharmcol., 32:718 (1980).
10. P. Steele, M. Weily, J. Rainwater, and R. Vogel, Platelet survival time and thromboembolism in patients with mitral

valve prolapse, Circulation, 60:43-54 (1979).

11. M. Cortellaro, C. Boschetti, G. Fassio, M. Basagni, and E.E. Polli, A controlled study of the effect of sulfinpyrazone on platelet survival and on platelet-bound ^{14}C-serotonin release in patients with previous myocardial infarction, Acta Haematol., 61:68-74 (1979).

12. S. Moncada and R. Korbut, Dipyridamole and other phospho-diesterase inhibitors act as antithrombotic agents by potentiating endogenous prostacyclin, Lancet, i:1286-1289 (1978).

13. A. Maseri, Pathogenesis of the clinical events in ischemic heart disease, in: "Secondary Prevention of Ischemic Cardiac Events: Present Status and New Perspectives," E.E. Polli and M. Cortellaro, eds., Hans Huber, Bern (1983).

14. L.A. Cobb, R.S. Baum, H. Alvarez III, and W.A. Schaffer, Resuscitation from out-of-hospital ventricular fibrillation: 4 years follow-up, Circulation, 52(Suppl. III):111-223 (1975).

15. M.A. De Wood, J. Spores, and R. Notske, Prevalence of total coronary occlusion during the early hours of transmural myocardial infarction, N. Engl. J. Med., 303:897-902 (1980).

16. R.L. Ridolfi and G.M. Hutchinson, The relationship between coronary artery lesions and myocardial infarcts: ulceration of atherosclerotic plaques precipitating coronary thrombo-sis, Am. Heart J., 93:468-486 (1977).

17. Anturane Reinfarction Trial Research Group, Sulfinpyrazone in the prevention of sudden death after myocardial infarc-tion, N. Engl. J. Med., 302:250-256 (1980).

18. Anturan Reinfarction Italian Study, Sulfinpyrazone in post-myocardial infarction, Lancet, i:237-242 (1982).

19. The FDA's critique of the Anturane Reinfarction Trial, N. Engl. J. Med., 303:1488-1492 (1980).

20. P.C. Elwood, A.L. Cochrane, and M.L. Burr, A randomized controlled trial of acetylsalicylic acid in the secondary prevention of mortality from myocardial infarction, Br. Med. J., 1:436 (1974).

21. P.C. Elwood and P.M. Sweetman, Aspirin and secondary mortality after myocardial infarction, Lancet, ii:1313 (1979).

22. Coronary Drug Project Research Group, Aspirin in coronary heart disease, J. Clin. Dis., 29:625 (1976).

23. K. Breddin, German-Austrian multicentre prospective study on

the prevention of secondary myocardial infarction by aspirin in comparison to phenprocoumon and placebo, Thromb. Haemost., 38:168 (1977).

24. Aspirin Myocardial Infarction Study Research Group, A randomized controlled trial of aspirin in persons recovered from myocardial infarction, J.A.M.A., 243:661 (1980).

25. Persantine-Aspirin Re-Infarction Study Research Group, Persantine and aspirin in coronary heart disease, Circulation, 62:449 (1980).

26. The E.P.S.I.M. Research Group, A controlled comparison of aspirin and oral anticoagulants in prevention of death. after myocardial infarction, N. Engl. J. Med., 307:701-708 (1982).

27. M. Moser, W. Plains, and R. Goslin, β-blockers and myocardial infarction, Arch. Intern. Med., 142:1618 (1982).

28. W.B. Campbell, A.R. Johnson, K.S. Callahan, and R.M. Graham, Anti-platelet activity of beta-adrenergic antagonists: inhibition of thromboxane synthesis and platelet aggregation in patients receiving long-term propanolol treatment, Lancet, ii:1382 (1981).

29. K.W. Hansen, R. Klysner, A. Geisler, J.B. Knudsen, S. Glarer, and J. Gormsen, Platelet aggregation and beta-blockers, Lancet, i:224 (1982).

HEMOSTATIC ALTERATIONS IN

DIABETES MELLITUS

Gian Franco Gensini

Università di Florence
Clinica Medica I
Head: Prof. G.G. Neri Serneri
Florence, Italy

INTRODUCTION

Cardiovacular complications are very common in patients with diabetes mellitus and play an important role in the natural history of diabetes. Therefore, many investigations were directed toward the study of the genesis of such complications. Platelets and blood clotting have been investigated in the light of their possible importance in the development of occlusive vascular disease. However, in spite of extensive investigations in this field, the well defined knowledge of hemostatic changes, and of their relationship with diabetes and its vascular complications is still relatively little. This probably depends on the fact that diabetes is still an heterogeneous syndrome resulting from a number of causes; moreover, its severe metabolic changes can by themselves affect platelet and clotting functions, and clinically undetectable complications may represent both a cause and an effect of hemostatic changes.

A series of reports indicate the occurrence of significant changes in platelet functions, in blood clotting, and in fibrinolysis in patients with diabetes mellitus. These functional changes are not related to age, sex, the duration and the type of diabetes (insulin-dependent or insulin-independent), to serum cholesterol, and serum triglyceride values (Hellem et al., 1964; Leone et al., 1974; Almer and Nilsson, 1975; Sagel et al., 1975; Fleischman et al., 1976; Gensini et al., 1979; Malhotra et al., 1982; Preston, 1982).

BLOOD PLATELETS

Platelet aggregation has been found to be enhanced in diabetes, especially in diabetes complicated by microvascular angiopathy (retinopathy, proteinuric nephropathy) or by atherosclerosis (Badawi et al., 1970; Bensoussan et al., 1975; Fleischman et al., 1976). A distinctive feature of platelets from diabetics is their increased susceptibility to various aggregating agents (ADP, adrenaline, and collagen – Colwell et al., 1976; Butkus et al., 1980). This is an early change in the natural history of diabetes, because subjects with impaired glucose tolerance show a significant increase in the so-called "circulating platelet aggregates" (Gensini et al., 1979). The increased aggreation is associated with enhanced thrombocytopoiesis as indicated by the increased number of mega-thrombocytes, thus suggesting that in vivo platelet activation leads to platelet consumption (Neri Serneri et al., 1979; Tindall et al., 1981). Moreover, increased plasma levels of the platelet-specific protein beta-thromboglobulin, indicating in vivo platelet activation, have been found in diabetics (Burrows et al., 1978; Preston et al., 1978; Zahavi et al., 1979).

During recent years, the attention of the researchers has been mainly focused on the mechanisms underlying the enhanced platelet aggregation. Both intra- and extra-platelet mechanisms seem to be involved:

1. Platelets from patients with impaired glucose tolerance and with diabetes mellitus produce larger amounts of thromboxane A_2 than platelets from control subjects when stimulated with thrombin or arachidonic acid (Neri Serneri et al., 1979; Butkus et al., 1980; Ziboh et al., 1980; Halushka et al., 1981). The increased production of thromboxane seems to be related both to increased phospholipase activity of platelet membrane (Takeda et al., 1981), and to increased activity of thromboxane synthetase (Butkus et al., 1980).

2. Platelets from diabetic patients are less sensitive to the inhibitory activity of PGD_2 and PGI_2 than platelets from controls (Abbate et al., 1981; Davì et al., 1982).

3. In plasma from patients with diabetes an ill-defined factor has been described able to enhance platelet aggregation

(Kwaan et al., 1972) which could be related to the circulating insulin - anti-insulin immunocomplexes (Van Zile et al., 1981).

4. A platelet aggregating activity related to activated factor X and to von Willebrand factor has been found in plasma from diabetic patients (Neri Serneri et al., 1976,1980).

BLOOD CLOTTING

Changes in blood clotting have been detected in patients with diabetes with various methods. Fuller et al. (1979) reported increased factor VII activity, and fibrinogen concentration and increased coagulant activity of factor VIII have been found by Bern (1978), by Neri Serneri et al. (1979), and by Borkenstein et al. (1982). The indirect evidence of increased thrombin formation in vivo has been obtained in diabetic patients by the chromatographic assay of soluble fibrin complexes (Gensini et al., 1979; Tsianos and Stathakis, 1980), and by the radioimmunoassay of fibrinopeptide A (Neri Serneri et al., 1979). The elevated antithrombin III levels reported by Corbella et al. (1979) and by Fuller et al. (1979) have also been attributed to enhanced antithrombin III production in response to increased thrombin formation in vivo.

FIBRINOLYSIS

Impairment of fibrinolytic system activity has been reported. Almer and Nilsson (1975) found a decreased plasminogen activator content of vessel wall, and Fuller et al. (1979) reported a decreased plasminogen activator activity in plasma from diabetics.

VESSEL WALL

In diabetic patients, an increased plasma concentration of factor VIII R antigen has been found (Neri Serneri et al., 1979; Borkenstein et al., 1982; Giustolisi et al., 1982), thus suggesting the existence of endothelial damage. Moreover, a decreased production of prostacyclin by vessel wall has been observed in venous tissue from patients with juvenile-onset diabetes (Silberbauer et al., 1979). Prostacyclin-like activity, assessed by bioassay, has been found to be significantly lowered both in insulin-dependent and in insulin-independent diabetics (Neri Serneri et al., 1979).

CONCLUSION

All the different components of the hemostatic system seem to
undergo significant changes in patients with diabetes mellitus.
Platelet functions are markedly altered because of an irregularity
of both the intra- and extra-platelet control mechanisms. The
activation of blood clotting, and the depression of the fibrinolytic
system together with the functional alterations of the endothelium
could lead to a facilitation of the thrombotic processes. However,
in spite of these large series of changes in hemostatic function,
their importance for the genesis and the progression of the vascular
complications of diabetes is still to be clearly established.

REFERENCES

Abbate, R., Navalesi, R., Benzi, L., Fortini, A., Gensini, G.F.,
        and Neri Serneri, G.G., 1981, Platelet functions in patients
        at thrombotic risk, Italian National Research Council -
        Atherosclerosis Project Meeting, Rome.
Almer, L.O., and Nilsson, I.M., 1975, On Fibrinolysis in diabetes
        mellitus, Acta Med. Scand., 198:101.
Badawi, H., El-Sawy, M., and Mikhai, M., 1970, Platelets, coagula-
        tion and fibrinolysis in diabetic and non-diabetic patients
        with quiescent coronary heart disease, Angiology, 21:511.
Bensoussan, D., Levy-Toledano, S., and Passa, P., 1975, Platelet
        hyperaggregation and increased plasma level of von Willebrand
        factor in diabetics with retinopathy, Diabetologia, 11:307.
Bern, M.M., 1978, Platelet functions in diabetes mellitus, Diabetes,
        27:342.
Borkenstein, M.H., and Muntean, W.E., 1982, Elevated factor VIII
        activity and factor VIII-related antigen in diabetic children
        without vascular disease, Diabetes, 31:1006.
Burrows, A.W., Chavin, S.I., and Hockaday, T.D., 1978, Plasma beta-
        thromboglobulin in concentrations in diabetes mellitus,
        Lancet, i:235.
Butkus, A., Skrinska, V.A., and Schumaker, O.P., 1980, Thromboxane
        production and platelet aggregation in diabetic subjects with
        clinical complications, Thromb. Res., 19:211.
Colwell, A., Kalushka, P.V., Sarji, K.E., Levine, J., Sagel, J.,
        and Nair, R.M.G., 1976, Altered platelet function in diabetes
        mellitus, Diabetes, 25:826.
Corbella, E., Miragliotta, G., Masperi, R., Villa, S., Bini, A.,

de Gaetano, G., and Chiumello, G., 1979, Platelet aggregation and antithrombin III levels in diabetic children, Haemostasis, 8:30.

Davì, G., Rini, G.B., Averna, M., Novo, S., Di Fede, G., Pinto, A., Notarbartolo, A., and Strano, A., 1982, Thromboxane B_2 formation and platelet sensitivity to prostacyclin in insulin-dependent and insulin-independent diabetes, Thromb. Res., 26:359.

Fleischman, A.I., Marvin, L., Bierenbaum, A.S., Somol, S.L., and Watson, P.B., 1976, in vivo platelet function in diabetes mellitus, Thromb. Res., 9:467.

Fuller, J.H., Keen, H., Jarrett, R.J., Omer, T., Meade, T.W., Chakrabarti, R., North, V.R., and Stirling, Y., 1979, Hemostatic variables associated with diabetes and its complications, Br. Med. J., ii:964.

Gensini, G.F., Abbate, R., Favilla, S., and Neri Serneri, G.G., 1979, Changes of platelet function and blood clotting in diabetes mellitus, Thromb. Haemost., 42:983.

Giustolisi, R., Musso, R., Russo, M., Catania, N., Lombardo, T., and Cacciola, E., 1982, Possible evidence for an increased factor VIII antigen synthesis in vascular endothelium of diabetic subjects, Thromb. Haemost., 47:293.

Halushka, P.V., Rogers, R.C., Loadholt, C.B., and Colwell, J.A., 1981, Increased platelet thromboxane synthesis in diabetes mellitus, J. Lab. Clin. Med., 97:87.

Hellem, A.J., Skalhegg, B.A., and Odegard, A.E., 1964, The platelet adhesiveness in plasma from diabetic and lipemic patients, Sangre, 9:175.

Kwaan, H.C., Colwell, J.A., Cruz, S., Suwanwella, N., and Dobbie, J.G., 1972, Increased platelet aggregation in diabetes mellitus, J. Lab. Clin. Med., 80:236.

Leone, G., Bizzi, B., Accorrà, F., and Boni, P., 1974, Functional aspects of platelets in diabets mellitus, in: "Platelet Aggregation and Drugs," L. Caprino and E.C. Rossi, eds., Academic Press, New York, p. 49.

Malhotra, K.C., Arya, R.K., Chugh, S.N., and Singla, B.B., 1982, Status of platelets in complicated and uncomplicated diabetes mellitus, Angiology, 33:410.

Neri Serneri, G.G., Abbate, R., Gensini, G.F., and Mugnaini, C., 1976, Increased platelet aggregation due to a plasmatic aggregating activity (factor Xa?), Eur. Symp. Adv. Coagulation, Fibrinolysis, Platelet aggregation and Atherosclerosis, Palermo.

Neri Serneri, G.G., Abbate, R., Gensini, G.F., Masotti, G., Mannucci,
    P.M., and Strano, A., 1979, aterosclerosi e stati trombo-
    filici, Relazione all'80° Congr. Naz. Soc. It. Med. Interna,
    ed., Pozzi, Roma.

Neri Serneri, G.G., Abbate, R., Mugnaini, C., and Gensini, G.F.,
    1980, Increased platelet aggregation due to a plasma-
    aggregating activity. Identification of the responsible
    factors, Haemostasis, 9:141.

Preston, F.E., Ward, J.D., Marcola, B.H., Porter, N.R., Timperley,
    W.R., and O'Malley, B.C., 1978, Elevated beta-thromboglobulin
    levels and circulating platelet aggregates in diabetic
    microangiopathy, Lancet, i:238.

Preston, F.E., 1982, Disorders of haemostasis in diabetes mellitus,
    La Ricerca Clin. Lab., 12:425.

Sagel, J., Colwell, J.A., Crook, L., and Laimins, M., 1975, Increased
    platelet aggregation in early diabetes mellitus, Ann. Intern.
    Med., 82:733.

Silberbauer, K., Schernthaner, G., Sinzinger, H., Piza-Katzer, H.,
    and Winter, M., 1979, Decreased vascular prostacyclin in
    juvenile-onset diabetes, New Engl. J. Med., 300:366.

Takeda, H., Maeda, H., Fukushima, H., Nakamura, N., and Uzawa, H.,
    1981, Increased platelet phospholipase activity in diabetic
    subjects, Thromb. Res., 24:131.

Tindall, H., Paton, R.C., Zuzel, M., and McNicol, G.P., 1981,
    Platelet life span in diabetics with and without retinopathy,
    Thromb. Res., 21:641.

Tsianos, E.B., and Stathakis, N.E., 1980, Soluble fibrin complexes
    and fibrinogen heterogeneity in diabetes mellitus, Thromb.
    Haemost., 44:130.

Van Zile, J., Kilpatrick, M., Laimius, M., Sagel, J., Colwell,
    J.A., and Virella, G., 1981, Platelet aggregation and
    release of ATP after incubation with soluble immune complexes
    purified from the serum of diabetic patients, Diabetes,
    30:575.

Zahavi, J., Jones, N.A.G., Betteridge, D.J., Leyton, J., Galton,
    D.J., Clark, S.E., and Kakkar, V.V., 1979, Platelet factor
    4, beta-thromboglobulin, malondialdehyde formation and blood
    lipids in patients with diabetes mellitus, Thromb. Haemost.,
    42:334.

Ziboh, V., Maruta, H., Lord, J., Cagli, W.D., and Lucky, W., 1979,
    Increased biosynthesis of thromboxane A_2 by diabetic
    patients, Eur. J. Clin. Invest., 9:223.

HEMOSTATIC ALTERATIONS IN PERIPHERAL ARTERIOPATHIES

Antonio Strano, Giovanni Davì, Gino Avellone,
Salvatore Novo, and Antonio Pinto

Institute of Clinical Medicine
University of Palermo
Italy

INTRODUCTION

The most common cause of obliterative disease in limb arteries
is slowly progressive arteriosclerosis which is eventually superimposed by thrombosis.

Transient or long-standing abnormalities often occur as a
specific reflection of activation of blood coagulation, or the
platelet release reaction in association with peripheral vascular
disease. A systemic activation of blood coagulation may be caused
from interaction of plasma coagulation factors and platelets with
the injured vessel wall (1); if activation of blood coagulation
is sufficiently marked, thrombin is generated. Thrombin cleaves
fibrinopeptide A and B from fibrinogen, converting fibrinogen into
fibrin monomer (2). Through a feedback mechanism, it changes
factors V and VIII to a functional form, it stimulates prostaglandin
synthesis by platelets, and it induces the platelet release reaction
independently of prostaglandin synthesis (3).

Arterial thrombin usually occurs when platelets come in contact
with exposed subendothelium (4). The platelets adhere, undergo the
release reaction, and aggregate. If the aggregates are sufficiently
large or the atherosclerotic stenosis marked, an occlusive thrombus
is produced or, more frequently, the aggregates embolize to obstruct

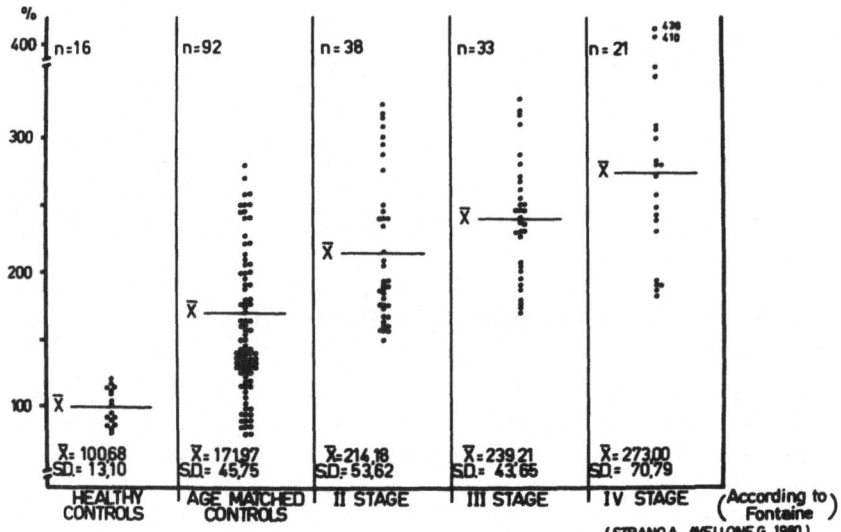

Fig. 1.   Factor VIII R:Ag in patients with peripheral
          vascular disease and controls.

the arterial system distally. To investigate the application of
platelet and coagulative tests in vascular disease, we studied
patients with peripheral vascular disease.

MATERIAL AND METHODS

     Two groups of patients (A and B) were studied.

Group A

     F. VIII R:Ag, F. VIII R:vW, F. VIII R:C, and biological
activity of antithrombin III (At III) were determined in 92 patients
(61 males and 31 females) with chronic peripheral vascular disease
due to arteriosclerosis. Diagnosis and classification (II, III,
and IV stage) were based on clinical examination and on strain-
gauge (Periflow-Janssen) plethysmography. Thirty-eight patients
(aged 40 to 58 years) were in the II stage group, 33 patients
(aged 51 to 70 years) in the III stage group, 21 patients (aged
64 to 76 years) in the IV stage group. Healthy controls of
equivalent age and sex were matched to the above patients.
Pharmacological treatment was suspended or reduced to the use of
analgesics only in the immediate days before blood sampling (3.5
days on the average).

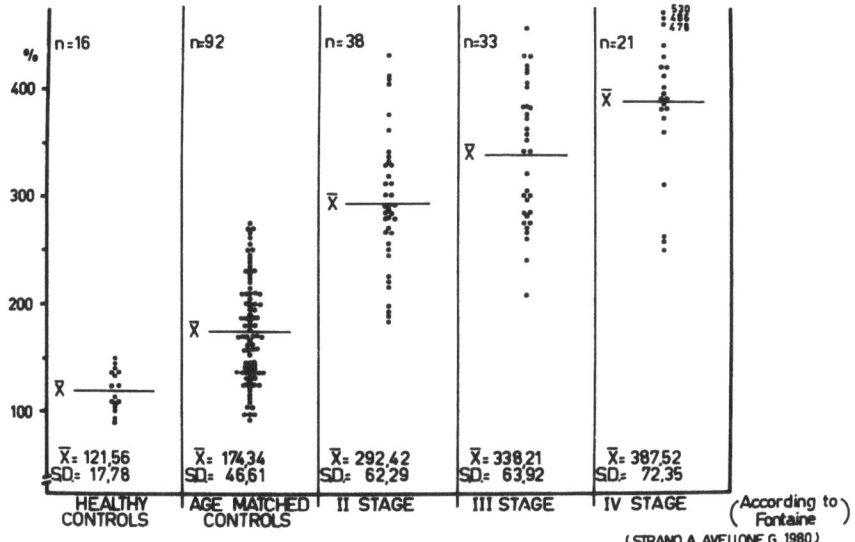

Fig. 2. Factor VIII R:C in patients with peripheral
          vascular disease and controls.

Blood was sampled after a fasting period of 12 hours with
atraumatic venipuncture, without stasis and with the double syringe
technique. All the samples were taken with polypropilene syringes
containing 1/10th volume of sodium citrate 0.11 mol/1. All specimens
were assayed within 4 hours of blood sampling.

Factor VIII R:C was measured using the two time method (5),
factor VIII R:Ag using immunoelectrophoresis (6), factor VIII R:vW
with Weiss' method (7), preparing the platelet suspension with
platelets washed according to Walsh's technique (8). Antithrombin
III has been determined as antithrombin activity according to
von Kaulla's method (9) on serum obtained from 3 ml of whole blood
set in a glass test tube and left to coagulate at room temperature
(10,11).

Group B

Plasma β-thromboglobulin (βTG), platelet aggregatio ratio (PAR),
platelet regeneration time (PRT), and thromboxane B_2 (TXB_2) formation by platelets were determined in 10 PVD patients (7 males and
3 females) ranging in age from 48 to 69 years. The sex
and the age were matched with 10 apparently healthy controls.

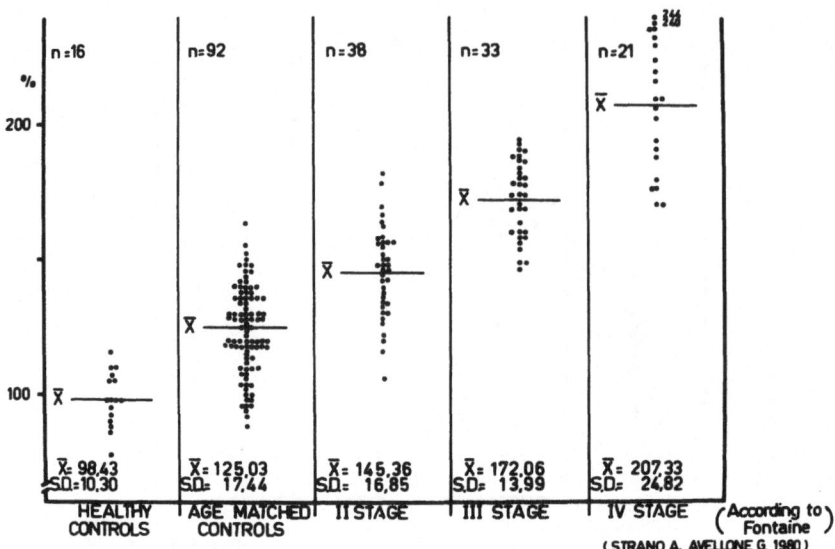

Fig. 3.   Factor VIII R:vW in patients with peripheral
          vascular disease and controls.

Diagnosis of peripheral vascular disease, was based on clinical
examination and on strain-gauge plethysmography. Neither the
controls nor the diabetic patients had taken aspirin or any other
antiplatelet agent for at least 2 weeks prior to investigation.

Blood samples were collected from the antecubital vein of
subjects fasted. βTG assay was performed on venous blood sample
drawn without stasis into polypropilene syringes containing platelet
release inhibitors (PGE$_1$, EDTA, and Theophylline). Tubes were immediately placed in crushed ice. Plasma was separated within 2
hours from collection by centrifugation at 4°C. The middle third
of the platelet-poor plasma was removed for βTG radioimmunoassay,
as previously described (12,13).

Thromboxane B$_2$ levels were measured in samples of platelet-
rich plasma under conditions similar to those used for aggregation
studies. Two hundred μl of thrombin (5 U/ml) for 4 minutes. TXB$_2$
content was measured in appropriate dilutions of supernatant using
a RIA technique (14,15,16). Platelet aggregatio ratio was determined
before and after bicycle ergometer test as previously described
(17,18). Platelet regeneration time was studied as previously
described (19,20).

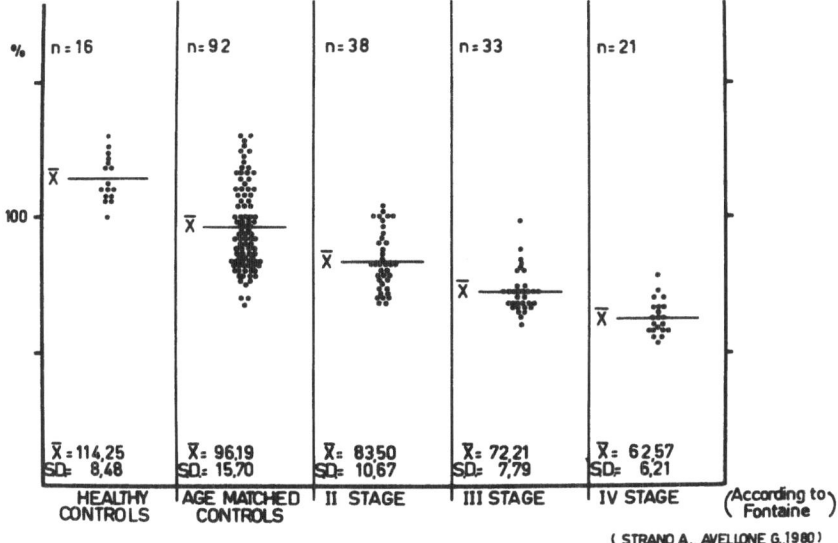

Fig. 4.   % AT III activity in patients with peripheral
          vascular disease and controls.

Statistical Analysis

Differences between the means have been evaluated with
Student's t test. All data presented as the mean $\pm$ S.D.

RESULTS

Factor VIII R:Ag (Fig. 1) was increased in stage II patients
(214.18 $\pm$ 53.62%), in stage III patients (239.21 $\pm$ 43.65%), and in
stage IV patients (273 $\pm$ 70.79%) in comparison to the matched
controls (171.97 $\pm$ 45.75%). A similar result was found for factor
VIII R:C (Fig. 2) and factor VIII R:vW (Fig. 3), showing a progressive increase of activity with the progression of the disease.

These changes are linked to a thrombin-like mechanism made
evident by a costant reduction of the biological activity of antithrombin III, which decreases progressively with the increasing
stage of the arteriopathy (Fig. 4).

A significant regression coefficient between factor VIII R:C
and antithrombin III (r = -0.57; p<0.005) was observed (Fig. 5);
plasma βTG levels (Fig. 6) were also increased (47.5 $\pm$ 21.5 ng/ml)
in PVD patients in comparison to matched controls (19.2 $\pm$ 8.6 ng/ml).

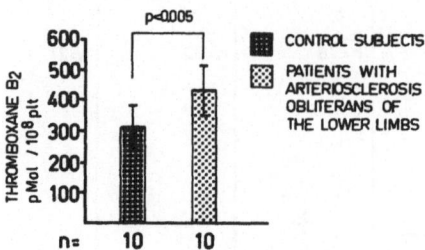

Fig. 5.   TXB$_2$ formation
in PRP after
thrombin (5 Ux4
min) stimulation
in patients with
peripheral vas-
cular disease
and controls.

An evident reduction of PRT in PVD patients was also observed ($t\frac{1}{2}$ 2.6 $\pm$ 0.8 days in PVD patients and 3.5 $\pm$ 0.6 in matched controls) (Fig. 7).

Figure 8 shows the changes obtained after exercise in PAR and in the PVD patients and in the matched controls.

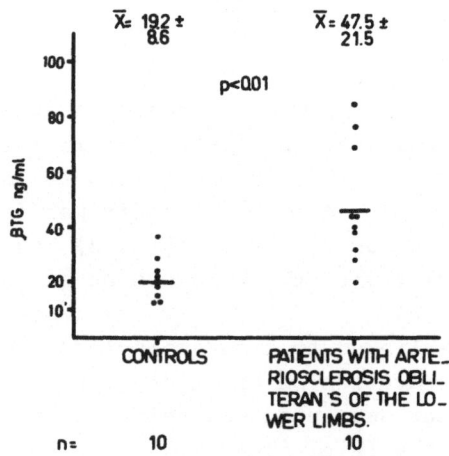

Fig. 6.   Plasma βTG levels
in patients with
peripheral vas-
cular disease and
controls.

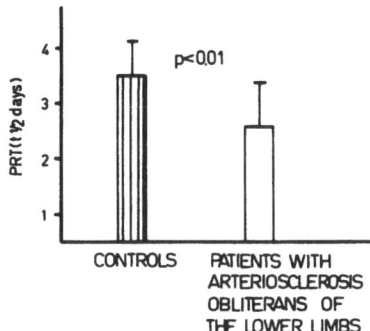

Fig. 7. Platelet regeneration time in
patients with
peripheral vas-
cular disease
and controls.

DISCUSSION

The increased mortality in claudicating patients corresponds
to about the mortality rate one expects in an overall population
10 years older; that could be, in fact, explained by the coagulative
and platelet activation frequently found in PVD patients.

In fact there are changes of the coagulation system showing
an acceleration of the enzymatic processes that lead to the formation of thrombin. The increase of factor VIII R:C and factor VIII
R:vW (subcomponents of factor VIII sensitive to thrombin), and
simultaneous reduction of antithrombin III activity confirm thrombin
formation caused by an activation of coagulation factors above
factor Xa of the intrinsic pathway, partially inhibited by physiological inhibitor. Moreover, the long compensation phase that
precedes an acute thromboembolic episode is linked to other
plasmatic and to tissue components able to neutralize the effects
of thrombin. The marked decrease of AT III in stage IV patients
could be expression of a prethrombotic situation in the micro-
circle district.

In agreement with other investigators, we found that factor
VIII R:Ag was significantly elevated in PVD patients (21,22).
Plasma β-thromboglobulin was significantly increased in PVD
patients in agreement with recent data of Cella et al. (23)

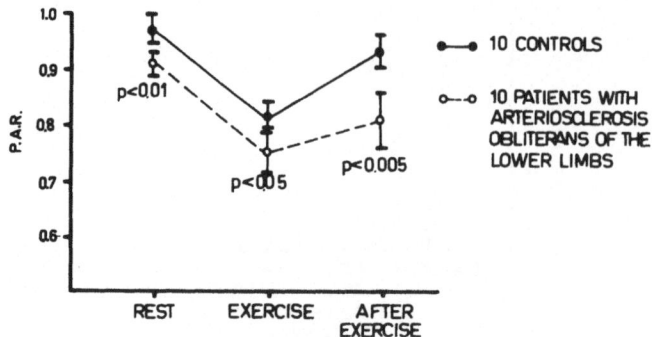

Fig. 8. Platelet aggregatio ratio before
         and after exercise in patients
         with peripheral vascular dis-
         ease and controls.

and Baele et al. (24). There was no correlation between plasma
βTG after treadmill exercise till occurrence of claudicatio points
to platelet activation in vivo. A similar result was observed in
our study with the circulating platelet aggregate ratio. Platelets
may participate in the development of exercise-induced ischemia
in the lower limbs by forming flow-limiting platelet plugs on
atherosclerotic plaques. Platelet activation may also be secondary
to changes in flow dynamic; in fact, in areas of high turbulence
some platelet damage may occur.

The augmented TXB_2 formation and the increased platelet
turnover, showed in our study by reduced platelet regeneration
time, could indicate a persistent platelet activation with increased platelet consumption.

It is, therefore, important to know whether antiplatelet
drugs would reduce platelet activation and, would prevent platelet
deposition on damaged endothelium, and might simultaneously reduce
fibrin deposition and retard the progression of arteriosclerosis.

REFERNCES

1. J. Hirsh and E. Genton, Thrombogenesis, in: "Physiological
       Pharmacology," W.S. Root and N.L. Berlin, eds., Academic
       Press, New York (1974), pp. 99-133.
2. H.L. Nossel, M. Ti, K.L. Kaplan, K. Spanondis, T. Soland, and
       V.P. Butler, Jr., The generation of fibrinopeptide A in

clinical blood samples, J. Clin. Invest., 58:1136 (1976).

3. A.I. Schafer and R.I. Handin, The role of platelets in thrombotic and vascular disease, Prog. Cardiovasc. Dis., 22:31 (1979).

4. J.F. Mustard and M.A. Packham, The role of blood and platelets in atherosclerosis and the complications of atherosclerosis, Thromb. Diath. Haemorrh., 33:444 (1975).

5. K.W.E. Denson, in: "Human Blood Coagulation, Haemostasis and Thrombosis," R. Biggs, ed., Blackwell Scienfic Publication, Oxford-Edinburgh (1976).

6. C.B. Laurell, Electroimmunoassay, Scand. J. Clin. Lab. Invest., 29 Suppl. 124:21 (1972).

7. H.J. Weiss, L.W. Hoyer, R.F. Rickles, A. Varma, and J. Rogers, Quantitative assay of a plasma factor deficient of von Willebrand disease that is necessary for platelet aggregation: relationship to decreased F. VIII procoagulant activity and antigen content, J. Clin. Invest., 52:2708 (1973).

8. P.N. Walsh, Platelet coagulant activities and haemostasis: a hypothesis, Blood, 43:597 (1974).

9. E. Von Kaulla and K.N. Von Kaulla, Antithrombin III and disease, Am. J. Clin. Pathol., 48:69 (1967).

10. G. Avellone, V. Mandalà, F. Riolo, and A. Pinto, AT III in patients with atherosclerotic disease, in: "Vascular Occlusion: Epidemiological, Pathophysiological, and Therapeutic Aspects," Serono Symp. 37, M. Tesi and J. Dormandy, eds., Academic Press, New York, (1981), pp. 193-197.

11. G. Avellone, G. Davì, A. Pinto, and V. Mandalà, L'antitrombina III nella cardiopatia ischemica, Boll. Sic., XXIV, 2, 127-133 (1979).

12. G. Davì, G.B. Rini, M. Averna, S. Novo, G. Di Fede, A. Mattina, A. Notarbartolo, and A. Strano, Enhanced platelet release reaction in insulin-dependent and insulin-independent diabetic patients, Haemostasis, 12:275-281 (1982).

13. G. Davì, S. Novo, A. Pinto, N. Custro, M. Averna, A. Mattina, and A. Strano, Platelet activation after adrenergic stimulation in hypertensive patients: effects of acebutolol, Eur. Heart J., 4:295-299 (1983).

14. A. Strano, G. Davì, M. Averna, G.B. Rini, S. Novo, G. Di Fede, A. Mattina, and A. Notarbartolo, Platelet sensitivity to prostacyclin and thromboxane production in hyperlipidemic patients, Thromb. Haemost., 48:18-20 (1982).

15. A. Strano, G. Davì, M. Traina, S. Novo, and A. Ranieri,
     Thromboxane formation by platelets and platelet sensitivity
     to prostacyclin in patients with acute myocardial infarc-
     tion, Thromb. Haemost., 46:759 (1981).

16. G. Davì, G.B. Rini, M. Averna, S. Novo, G. Di Fede, A. Pinto,
     A. Notarbartolo, and A. Strano, Thromboxane B_2 formation
     and platelet sensitivity to prostacyclin in insulin-
     dependent and insulin-independent diabetics, Thromb. Res.,
     26:359-370 (1982).

17. K.K. Wu and J.C. Hoak, A new method for the quantitative
     detection of platelet aggregates in patients with arterial
     insufficiency, Lancet, 2:924 (1974).

18. A. Strano, S. Novo, G. Davì, G. Avellone, and A. Pinto,
     Platelet aggregation coagulation and fibrinolysis at rest
     and after bicycle ergometer test in CHD, in: "Selected
     Topics in Exercise Cardiology and Rehabilitation," A.
     Raineri, J.J. Kellermann, and V. Rulli, eds., Plenum
     Press, London, (1980). pp. 265-276.

19. A. Strano, S. Novo, G. Davì, G. Avellone, and V. Mandalà,
     Effects of sorbinicate and nicotinic acid on blood vis-
     cosity, red cell deformation and platelet function,
     Pharm. Res. Comm., 14:639-648 (1982).

20. G. Davì, G.B. Rini, M. Averna, S. Novo, G. Di Fede, A. Mattina,
     M. Fiore, and A. Notarbartolo, Misurazione del tempo di
     rigenerazione piastrinica in pazienti dislipidemici,
     Rassegna di Medicina Interna, 3:41-46 (1982).

21. H.C. Kwaan, J.A. Colwell, S. Cruz, N. Suwanwella, and J.G.
     Dobbie, Increased platelet aggregation in diabetes
     mellitus, J. Lab. Clin. Med., 80:236-246 (1972).

22. M.M. Bern, M.P. Cassani, J. Horton, L. Rand, and G. Davis,
     Changes in fibrinolysis and factor VIII coagulant, antigen,
     and ristocetin cofactor in diabetes mellitus and athero-
     sclerosis, Thromb. Res., 19:831-839 (1980).

23. G. Cella, J. Zahavi, H.A. de Haas, and V.V. Kakkar, β-thrombo-
     globulin, platelet production time, and platelet function
     in vascular disease, Br. J. Haematol., 43:127-136 (1979).

24. G. Baele, H. Bogaerts, D.L. Clement, R. Pannier, and F.
     Barbier, Platelet activation during treadmill exercise
     in patients with chronic peripheral arterial disease,
     Thromb. Res., 23:215 (1981).

HEMOSTATIC ALTERATIONS IN

CEREBROVASCULAR INSUFFICIENCY

M. Prencipe, C. Fieschi, S. Paolucci,
and C. Buttinelli

Clinica Neurologica I
Università di Roma

INTRODUCTION

Understanding of hemostatic alterations is relevant to the problem of treatment of TIAs and prevention of stroke. In fact, not all of cerebrovascular pathology is caused by embolic or atherothrombotic processes, and a single treatment may not necessarily be effective for all categories of TIA patients. The aim of studying platelet function is to try to identify patients potentially "responsive" and "unresponsive" to antiplatelet drugs, and to evaluate the strategies for treatment, by this meaning the drug(s) and drug schedule of choice. This will never be attained, however, without a preselection of patients based on identification of the pathogenetic mechanisms and the specific risk of stroke. For example, the inhibition of platelet thrombi by drugs or diets that interfere with any component of the atherothrombotic process would not necessarily be expected to influence the natural history of TIAs due to hemodynamic mechanisms, or due to advanced and severe diffuse disease of the cerebral arteries, or to embolism of calcific material from malfunctioning heart valves, nor perhaps to prevent embolization from thrombus within the heart chambers. Natural history of such subgroups of patients, ex vivo effect of antiplatelet drugs on hemostatic functions in the same, and perspective for enhanced pharmacological prevention of stroke will be the object of further studies.

In the present study we report our experience in using the method described by Wu and Hoak (1) for the detection of platelet aggregates.

METHOD

Platelet aggregates were detected by the method of Wu and Hoak (1) with minor modifications. One ml of freely flowing venous blood was drawn into each synringe: the first one containing 4 ml of edetic acid plus formalin, the second one 4 ml of edetic acid without formalin. Both samples were gently mixed and transferred to polypropylene tubes. After 15 minutes of incubation at room temperature, samples were centrifuged at 150 g for 8 minutes to prepare the platelet-rich plasma (PRP). Platelet count of each PRP was performed using a ZF Coulter Counter.

The number of platelet aggregates was expressed by the ratio between PRP with edetic acid plus formalin and PRP with edetic acid. All platelet aggregate ratio (P.A.R.) values have been multiplied by 100. Since platelet aggregates are thought to be fixed by formalin and then centrifuged out, low P.A.R. values indicate the presence of platelet aggregates.

PATIENTS

P.A.R. values have been evaluated in 318 subjects. None of them had taken aspirin-containing products or other non-steroidal antiinflammatory drugs in the previous 14 days. The cerebro-vascular group included 191 patients who had had at least one reversible ischemic attack (RIA). As normal and control groups, we studied 90 apparently healthy subjects, 24 patients with proven type II hyperlipoproteinemia and 13 diabetic patients. A careful history and clinical and laboratory examinations were performed to rule out symptomatic vascular diseases. Sex and age distribution of each group are reported in Table 1.

ANTIPLATELET THERAPIES

Seventy-nine patients with two consecutive (7 days interval) pathological P.A.R. values were submitted to different therapies: 25 patients to aspirin (325 mg daily), 20 to ditazole (1,200 mg daily), 24 to ticlopidine (500 mg daily), and 13 to a dietary

Table 1. P.A.R. Values in Different Pathological Conditions

	NUMBER OF SUBJECTS	MEAN AGE	SEX	P.A.R. VALUES MEAN $\pm$ S.D.	PATHOLOGICAL P.A.R. VALUES (%)
NORMAL SUBJECTS	90	45.1	21 F 69 M	95.4 $\pm$ 6.1	—
HYPERLIPIDAEMIC PATIENTS	24	46.2	11 F 13 M	82.9 $\pm$ 8.3[*]	50.0
DIABETIC PATIENTS	13	58.9	5 F 8 M	83.4 $\pm$ 8.5[*]	61.5
R.I.A. PATIENTS	191	58.1	71 F 120 M	76.3 $\pm$ 10.7[**]	81.0

STUDENT t-TEST: [*] P < 0.01   [**] P < 0.001

supplementation with polyunsaturated fatty acids (Naudicelle, 6 capsules daily). P.A.R. determinations were repeated after a week and a month of treatment. Ditazole, ticlopidine, and Naudicelle groups were also evaluated 30 days after the interruption of therapy. Lastly, P.A.R. determinations were performed in two groups of RIA patients chronically treated (at lest 30 days) with sulfinpyrazone (13 patients taking 800 mg daily) and dipyridamole (50 patients taking 225 mg daily). Baseline P.A.R. values of these two groups are not available.

RESULTS

Reproducibility

To test the reproducibility of the technique P.A.R. determinations were performed repeatedly in basal conditions. In 45 subjects the mean difference between P.A.R. values obtained from two blood samples drawn simultaneously was 0.04 with an SD of 3.23. Two samples drawn one week apart from 81 subjects showed slightly higher values (0.49 $\pm$ 4.32 SD). Three samples drawn at separate times from 36 subjects showed the following values: 0.08 $\pm$ 4.95 SD (1st vs. 2nd sample), 0.69 $\pm$ 4.10 SD (1st vs. 3rd sample), and 0.61 $\pm$ 5.54 SD (2nd vs. 3rd sample).

Table 2. P.A.R. Values in 191 R.I.A. Patients

DAYS FROM THE LAST ATTACK	NUMBER OF PATIENTS	P.A.R. VALUES MEAN ± S.D.	PATHOLOGICAL P.A.R. VALUES (%)
1 - 10	38	69.4 ± 9.6	97.4
11 - 30	80	77.4 ± 10.5*	81.2
31 - 365	73	77.0 ± 9.4*	81.0

STUDENT t-TEST vs BASELINE VALUE   *P < 0.001

Baseline P.A.R. Values

Mean P.A.R. value of normal subjects was 95.4 ± 6.1 SD without a significant difference between males (95.1 ± 6.5 SD) and females (95.7 ± 5.4 SD). P.A.R. values did not differ significantly among young (<30 years), middle-aged (30-60 years), and older healthy subjects (>60 years). Mean P.A.R. values of these three groups were respectively: 94.1 ± 6.6 SD, 95.9 ± 6.0 SD, and 94.3 ± 6.9 SD. Since the range of normal P.A.R. values varies from 86 to 108, we considered as pathological only values lower than 86.

As shown in Table 1, all groups of patients had P.A.R. values significantly lower than those of normal subjects: 50% of dyslipidemic patients, 61.5% of diabetic patients, and 81% of RIA patients had P.A.R. values lower than 86. The lowest mean P.A.R. value (76.3 ± 10.7 SD) was observed in the RIA group.

As shown in Table 2, 38 RIA patients studied within 10 days from the last ischemic attack had a mean P.A.R. value (69.4 ± 9.6 SD) significantly lower (p<0.001) than that of 80 patients from the 11th to the 30th day (77.4 ± 10.5 SD), and that of 73 patients studied from the 31st to the 365th day (77.0 ± 9.4 SD).

Amongst RIA patients no significant differences were noted according to the severity of neurological signs, or the presence of associated diseases such as hypertension, dyslipidemia, diabetes, and myocardial infarction; 41.9% of 80 patients without any risk factors showed pathological P.A.R. values.

Table 3. P.A.R. Mean Values + S.D.

		BASELINE	7 DAYS OF THERAPY	30 DAYS OF THERAPY	30 DAYS AFTER THERAPY
GROUP A	(A.S.A.) (No. 25)	71.0 +9.1	92.1*** +6.3 (88)	93.6 +3.5 (96)	————
GROUP B	(Ticlopidine) (No. 24)	73.0 +6.7	85.0*** +7.7 (43.5)	90.8** +9.6 (80)	80.6*** +7.0 (61.9)
GROUP C	(Ditazole) (No. 20)	73.8 +6.4	84.0*** +10.0 (45)	89.3* +6.0 (85)	78.3*** +5.4 (23)
GROUP D	(Naudicelle) (No. 13)	71.2 +8.9	79.4* +6.1 (16)	91.9*** +3.6 (92.3)	75.1*** +6.6 (16)

STUDENT t-TEST VS PREVIOUS VALUE   ***$P < 0.001$ **$P < 0.01$ *$P < 0.05$

IN PARANTHESES PER CENT OF NORMAL P.A.R. VALUES

Antiplatelet Therapies

As shown in Table 3, after a week of treatment a significant increase of P.A.R. values was observed in all groups: 88% of aspirin, 43.5% of ticlopidine, 45% of ditazole, and 16% of Naudicelle patients reached the normal range.

After 30 days of therapy, aspirin patients presented a further, but not significant, increase of mean P.A.R. value. Conversely, a significant additional increase was observed in all other groups. At that time, 96% of aspirin, 80% of ticlopidine, 85% of ditazole, 92.3% of Naudicelle patients had normal P.A.R. values. Thirty days after drug interruption, a significant decrease ($p < 0.001$) was observed in the Naudicelle, ticlopidine, and ditazole groups. This aspect was not tested in the aspirin group.

Lastly, patients chronically treated with sulfinpyrazone or dipyridamole showed a lower percentage (46% and 38%) of patients with normal P.A.R. values. Mean P.A.R. values of these two groups were respectively 81.8 + 8.8 SD and 81.2 + 6.8 SD.

COMMENT

Wu and Hoak (1) postulated that their method was able to detect circulating platelet aggregates, but others (2,3) have suggested that platelet aggregates are loosely formed during blood samples procedures. Since β-TG plasma levels are considered a specific marker of in vivo platelet release (4), a comparison between these two tests could solve the problem. Chen and Wu (5) reported a good correlation between P.A.R. and β-TG values, but 5 of 37 specimens showed extremely high β-TG values, and, without simultaneous platelet factor 4 determination, it is difficult to rule out a possible in vitro β-TG release (6).

Therefore, it is still unclear whether pathological P.A.R. values indicate that platelets are prone to form aggregates in the circulation, or during blood samples procedures.

Despite this uncertainty on the biological meaning of the test, our data indicate that P.A.R. measurements have a good reproducibility. From the clinical point of view, P.A.R. values of RIA patients resulted significantly lower than those of normal subjects. As it has been suggested by other investigators (7), this platelet activation could be the consequence of the acute event; as a matter of fact, the group of patients studied shortly after the last ischemic attack showed the lowest mean P.A.R. value. However, pathological P.A.R. values were found in 81% of RIA patients studied a long time after the last attack, and no differences had been noted according to the severity of the neurological signs. In addition, P.A.R. values did not significantly differ in RIA patients with and without the same risk factors. Lastly, 41% of patients without any risk factors showed pathological P.A.R. values. Therefore, P.A.R. data suggest that RIA patients have a platelet hyperaggregability regardless of the presence of associated diseases and the length of time elapsed since the stroke occurred.

These findings are in agreement with results obtained with this and other methods which measure platelet adhesiveness, spontaneous or ADP-induced aggregation, platelet survival time, and β-TG plasma levels (8-12).

Unfortunately, at present no group has been prospectively followed to establish a casual relationship between platelet abnormalities and progress of the atherosclerotic process. The linkage
between platelet abnormalities and risk of thromboembolic events
in the cerebral circulation is even more intricate. First of all,
an ischemic cerebral lesion can be caused by mechanisms different
from the thromboembolic ones. In addition, a thromboembolic event
can be caused by material different from platelet aggregates (i.e.,
a mixture of cholesterol and atheromatous debris). Lastly, local
factors, namely seat and size of atherosclerotic plaques, play a
crucial role in the determination of cerebral embolic events.

Therefore, platelet abnormalities detected in the venous blood
are unlikely to be specific predictors of cerebral ischemic events
in a single subject. However, the comparison between groups of
cerebrovascular patients and normal people suggests that an abnormal
behavior of platelets, if repeatedly observed, should be associated
with an increased risk of thromboembolic events.

Lastly, platelet tests can be employed to demonstrate the
biological efficacy of antiplatelet drugs. P.A.R. modifications
observed during and after different antiplatelet therapies are a
further evidence for the validity of the test. Unfortunately,
the clinical usefulness of this and other platelet tests is still
lacking. Only long-term trials could indicate whether a correlation
exists between positive clinical effects of antiplatelet therapy
and normalization of platelet tests. However, the clinical effect
of aspirin is significantly lower than that observed on P.A.R.
and on other platelet tests (nearly 100% of normalization). This
fact suggests that platelet tests, at least those influenced by
a block of arachidonic acid pathways, are not sufficient to predict
the efficacy of treatment.

REFERENCES

1.   K.K. Wu and J.C. Hoak, A new method for the quantitative
         detection of platelet aggregates in patients with arterial
         insufficiency, Lancet, ii:924 (1974).
2.   J.A. Prazich, S.I. Rapaport, J.R. Samples, and R. Engler,
         Platelet aggregate ratio. Standardization of technique

and tests results in patients with myocardial ischemia and patients with cerebrovascular disease, Thromb. Haemostas., 38:597 (1977).

3.   T.F. Roher, B. Pfister, C. Weber, P.R. Imhof, and P. Stucki, Validity of the Wu and Hoak method for the quantitative determination of platelet aggregation in vivo, Blut, 36:15 (1978).

4.   J. Zahavi and V.V. Kakkar, β-thromboglobulin. A specific marker of in vivo platelet release reaction, Thromb. Haemost., 44:23 (1980).

5.   Y.C. Chen and K.K. Wu, A comparison of methods for the study of platelet hyperfunction in thromboembolic disorders, Br. J. Haematol., 46:263 (1980).

6.   K.L. Kaplan and J. Owen, Plasma levels of β-thromboglobulin and platelet factor 4 as indices of platelet activation in vivo, Blood, 57:199 (1981).

7.   J.H. Dougherty, D.E. Levy, and B.B. Weksler, Platelet activation in acute cerebral ischemia. Serial measurements of platelet function in cerebral vascular disease, Lancet, i:821 (1977).

8.   K.K. Wu and J.C. Hoak, Increased platelet aggregates in patients with transient ischemic attacks, Stroke, 6:521 (1975).

9.   G.D.O. Lowe, M.M. Reavey, R.V. Johnston, C.D. Forbes, and C.R.M. Prentice, Increased platelet aggregates in vascular and non-vascular illness: correlation with plasma fibrinogen and effect of ancrod, Thromb. Res., 14:377 (1979).

10. M. Prencipe, S. Paolucci, C. Buttinelli, and C. Fieschi, Increased platelet aggregates in patients with transient ischemic attacks, in: "Cerebral Microcirculation and Metabolism," J. Cervos-Navarro and E. Fritshka, eds., Raven Press, New York (1981).

11. P.E. Hansen, J.H. Hansen, and S. Stenbjerg, Platelet aggregation in focal cerebral ischemia. A clinical study, Acta Neurol. Scand., 65:212 (1982).

12. J.A. Acheson, G. Danta, and E.C. Acheson, Platelet adhesiveness in patients with cerebrovascular disease, Atherosclerosis, 15:123 (1972).

MINOR FIBRINOLYTICS AND THROMBOLYTICS

IN CLINIC

M. Tesi

Division of Angiology
Ospedale di S. Maria Nuova
Florence, Italy

CLINICAL PATHOGENETIC CONSIDERATIONS

When angiopathic patients arrive in the hospital, they usually present symptomatologies caused by vascular occlusion with stoppage of flow or advanced stenosis with significant reduction of flow. This results in ischemia in the arterial region and stasis in the venous region. Thus total or partial vascular occlusions, almost always thrombotic (sometimes embolic for the arteries), comprise the essence of angiopathies, and represent 90% of the cases that come under our observation.

If vascular occlusions are the basis of angiologic pathology, therapy must be the recanalization of the obstructed vessels. The goal can be attained medically with thrombolytic treatment, or surgically by surgical recanalization. This is the rationale that justifies the use of fibrinolytics for physicians, and the use of antithrombotic treatment with anticoagulants, platelet antiaggregants, minor fibrinolytics, and rheological drugs.

A brief schematic representation of the diseases which comprise the majority of angiopathies in the angiological wards is shown in Table 1.

We can now ask what should treatment be for the main clinical conditions shown on the chart, and in particular those accompanied by vascular occlusions?

Table 1.   The Most Frequent Diseases in the Angiological Divisions

---

Arteriopathy: <u>Acute</u>   – Embolic
                          – Thrombotic

              <u>Chronic</u>   – Thrombotic
                          (thrombus on fibrous plaques during
                          atherosclerotic disease)

Phlebopathy: <u>Superficial</u> – Thrombophlebitis
                          (thrombus that usually appears in
                          varicose disease)
                        – Ulcers due to phlebostasis

              <u>Deep</u>   – Deep venous thrombosis (thrombus in
                          various conditions)
                        – Post thrombotic disease

Acute Embolic Arteriopathy

We believe that treatment must be surgical in arterial embolism.
In fact, surgical recanalization performed with a Fogarthy catheter
is quite simple, fast, and generally does not involve risks.
Thrombolytic treatment, carried out with urokinase (UK) can be used
only in cases of distal embolization where a catheter cannot be
used due to the reduced caliber of the vessels, or those rare
cases which cannot undergo surgery.

Acute Thrombotic Arteriopathy

We act differently in cases of acute arterial thrombosis.
The chosen treatment is medical which consists of pharmacological
disobliteration with UK.

Chronic Thrombotic Arteriopathy

There are no symptoms of the disease in the course of
atherosclerosis, where for years fibrous plaques spread through
the organism, and at times during the entire life of the patient.
The patients arrive at the clinic with signs of regional localization

Table 2. Chronic Atherosclerotic Arteriopathy with
Claudicatio Intermittens

Out-patients treated for 5 years°	Patients amputated in 5 years
1,300	21 (1.65%)

°Patients with claudicatio intermittens are never
hospitalized.

(cardiac, cerebral, limbs, and others) when on the intimal region
(fibrous plaque) fibrin is generated, that is the occluding or
stenosing thrombus is formed. Usually, the process is slow, the
thrombus is old, and there is no lysable fibrin; therefore, there
is no use for thrombolytic treatment. Medical or surgical treatment
is usually used and the results are generally good. Our group used
combinations of drugs, platelet antiaggregants, heparin anticoagulants, minor fibrinolytics, and rheological agents. The
percentage of amputations in our survey was definitely limited
both in claudicatio intermittens (Table 2) and in forms with
absolute insufficiency (Table 3).

Superficial Thrombophlebitis

Superficial thrombophlebitis usually appears during the course
of varicose diseases. The parietal responsability is dominant in
the pathogenesis of the disease. The thrombus adheres tightly
to the walls and there are no pulmonary embolisms. In these cases,
treatment with UK is not used because the patients do not come to
the hospital at the onset of the disease, the evolution of the disease is not disabling, and the thrombolytic therapy is complex and
expensive.

Table 3. Chronic Atherosclerotic Arteriopathy with Absolute
Insufficiency (Pain at Rest and often with Gangrene)

Patients hospitalized then followed-up for 5 years°	Patients amputated in 5 years
1,265	61 (5.50%)

°Patients with absolute insufficiency are always hospitalized.

Table 4. Fibrinolytic Drugs in Clinical Use

Minor Fibrinolytics or
Non-Thrombolytics   − Stanozolol per os
                                   − Acetylcholine in arterial
                                     perfusion
                                   − Mesoglycane sulphate,
                                     artery, vein, os
                                   − Penthosane polysulphate
                                     (SP54) per os

Major Fibrinolytics or
Thrombolytics   − Urokinase (UK)
                                   − Streptokinase (SK)
                                   − tissutal-Plasminogen
                                     Activator (t-PA)

Cortisonics and non-steroid anti-inflammatory drugs, heparin, antiaggregants, and minor fibrinolytics were used to act on the venous wall. Various localized treatment was also administered.

Deep Venous Thrombosis

In this instance, thrombogenesis is linked to hematic changes, due to the stasis and not to the alterations in the walls. As a result of the venous stasis of different duration, local hyper-coagulability with the activation of factor X and thrombin, platelet hyperactivity and appearence of TXA_2, hyperviscosity, and finally hypofibrinolysis with local breakdown of t-PA appear there. Often long thrombi are formed which do not adhere to the walls along with embolizing capability. Thrombolytic treatment is the most ideal; however, it must be initiated not more than a certain number of days after the onset of the disease.

THE DRUGS

There are numerous problems with fibrinolytic drugs, even if from a clinical point of view the fundamental question is that of their use with vascular patients. Therefore, in this second part of the report, which will deal with the pharmacological aspect, we shall approach the matter clinically in relation to our experience in the hospital.

Table 5. Main Pharmacological Features of the Two Groups of Drugs

Minor Fibrinolytics or Non-Thrombolytics	- Do not lyse to thrombi - Increase synthesis and secretion of t-PA - Activate local lysis mainly (+++) - Less activation of general lysis (+) - Increase hematic levels of anti-thrombin III - In part reduce platelet aggregation
Major Fibrinolytics or Thrombolytics	- Lyse the thrombi - Directly activate plasminogen to plasmin - Activate lysis inside the thrombus - UK and SK also activate hematic lysis - t-PA, while greatly activating lysis of the thrombus, does not really activate hematic lysis

The fibrinolytic drugs used in arteriopathy and phlebopathy can be grouped into minor fibrinolytics (non-thrombolytics) and major fibrinolytics (thrombolytics). As far as therapeutic activity is concerned, it is important to distinguish between drugs which can lyse the thrombi and those which are capable of inducing small local lysis of modest quantities of fibrin. The former are direct plasminogen activators; the latter are indirect activators of fibrinolysis, since they increase the production and secretion of t-PA by the vascular walls (Table 4).

From a clinical point of view, we must take into consideration some pharmacological features of the two groups of drugs (Table 5). These features are both indicated for use in acute forms where we tend to dissolve the existing thrombi, and in the chronic forms where it is useful to introduce a secondary prophylaxis which controls the growth of existing thrombi and the formation of new thrombi.

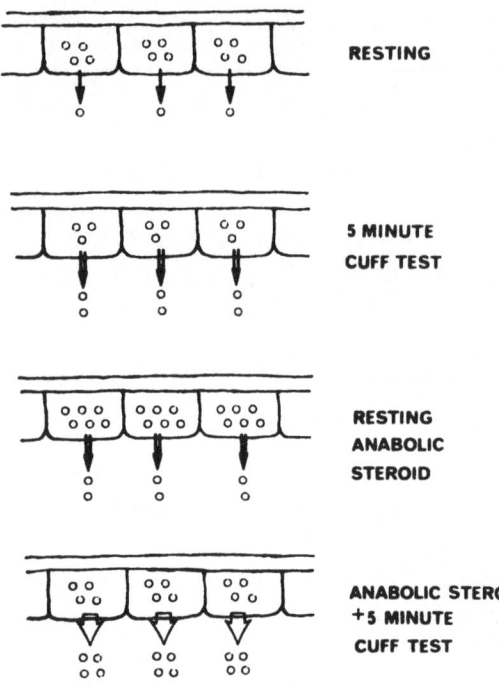

RESTING

5 MINUTE
CUFF TEST

RESTING
ANABOLIC
STEROID

ANABOLIC STEROID
+5 MINUTE
CUFF TEST

Fig. 1. Synthesis, storage, and release of plasminogen activator from the venous endothelium. In the resting state there is a steady rate of synthesis, a constant store, and a slow release. During a 5-min. cuff test the store is depleted and more activator is released into the circulation. After anabolic steroid therapy the rate of synthesis and the store are increased, as is the rate of release. After anabolic steroid a 5-min. cuff test releases increased amounts of activator from the increased stores.

Having summarized some of the main features of the main fibrinolytics in clinical use, we must now consider some of the properties specific to the individual drugs.

Stanozolol

We use this drug in conjuction with other antithrombotic drugs, such as heparin subcutaneous, antiaggregants such as ticlopidine and others, both in acute and in chronic forms. In particular, as far as the acute forms are concerned, it is used in thrombotic arterio-pathy, and deep venous thrombosis, after thrombolytic treatment, and also in the treatment of superficial thrombophlebitis. As for the chronic forms, it is used in the atherosclerotic arteriopathy and in post-thrombotic disease. The drug is administered orally, at a rate of 10 mg per day.

Background is provided in research, where increased production and secretion of t-PA by the endothelial cells of the vascular walls with stanozolol were observed (Davidson et al. 1975, Fig. 1).

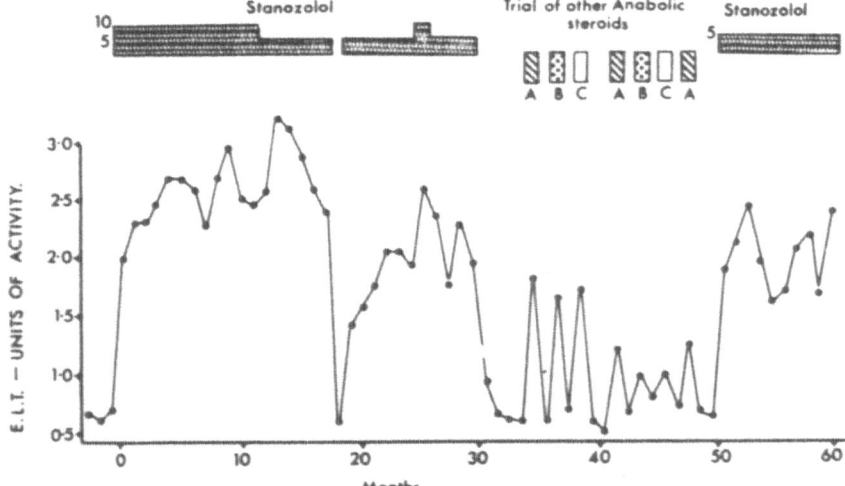

Fig. 2. Mean fibrinolytic activity, expressed as ELT
units, in all samples collected during baseline and monthly measurements (1 to 60 months).

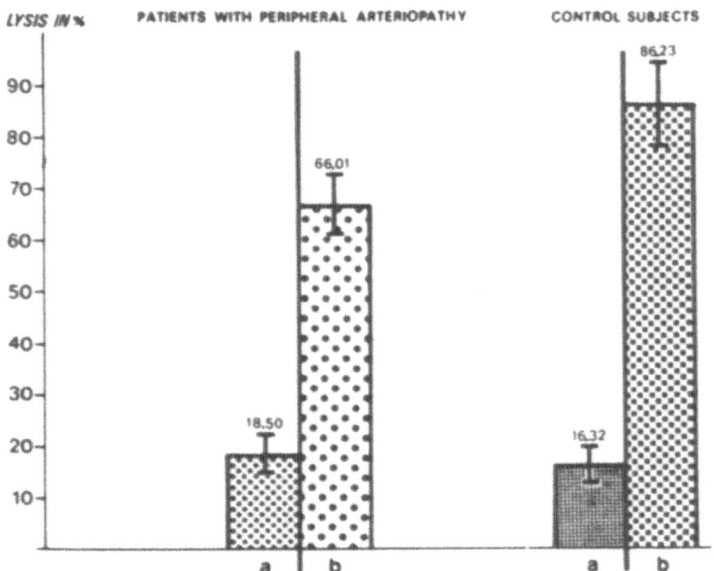

Fig. 3. Regional lysis. Zonal lysis of
the lower leg. Percentage lysis
in blood from femoral vein before
(a) and after (b) femoral arterial
injection of acetylcholine.

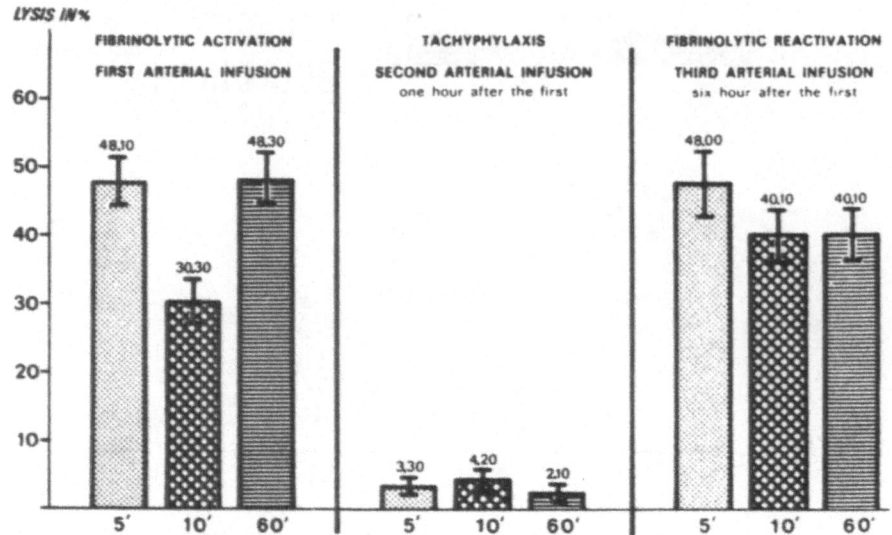

Fig. 4.   Regional tachyphylaxis in the lower leg zone.
         Percentage lysis in blood from the femoral
         vein during arterial infusion of acetylcholine
         in patients with peripheral arteriopathy.

The same group also observed that administration of the drug
is capable of activating lysis for months and even for some years
(Walker and Davidson, 1978; Fig. 2).

Acetylcholine

This drug was introduced in the treatment of peripheral vascular
diseases by our group (Andreotti and Nuzzaci, 1961; Pratesi et
al., 1962). The drug is indicated for chronic obliterating
arteriopathy of the atherosclerotic type and for endoarterial
infusions of 500 mg. From the evaluation of regional fibrinolysis
and through specimens taken from the femoral vein, before and
after arterial infusion of the drug in the affected limb, it was
possible to obtain good activation of lysis in the affected limb
as well (Tesi, 1975; Fig. 3).

Tachyphylaxis is a problem with fibrinolytic therapy. As far
as capability of activating lysis is concerned, 500 mg of acetylcholine infused in the femoral artery for 60 minutes activated
lysis for an entire hour. A second unfusion, one hour after the
completion of the first, did not create activation because of the

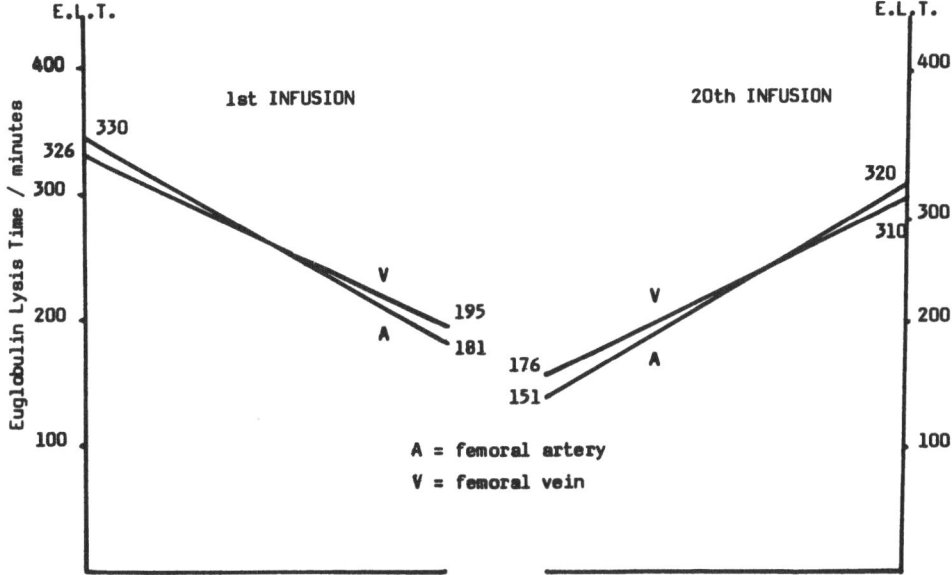

Fig. 5.   Euglobulin Lysis Time: test performed on the arterial
and venous blood of the lower occluded limbs. Differences betweeen before and after the 1st and 20th
arterial infusion of mesoglycane sulphate. Mean
values are expressed.

depletion of the regional fibrinolytic potential. If, however,
the second arterial infusion is administered 6 hours after the
first, the same lysis activation is obtained as in the first
infusion (Tesi, 1975; Fig. 4).

Mesoglycane Sulphate

A second drug studied by our group is a sufficiently purified
glycosoaminoglycan, since it is comprised of three electrophoretical-
ly identified fractions. The euglobulin lysis time, measured in
arterial and venous femoral blood in atherosclerotic obliterating
arteriopathy, before and after endoarterial infusion of 150 mg of
the drug, was significantly shortened both at the first and at
the twentieth infusion (Tesi et al., 1980; Fig. 5).

This means that even mesoglycane sulphate not only activates
lysis in general, but activates it in the arteriopathic patients,
and in their affected limb as well.

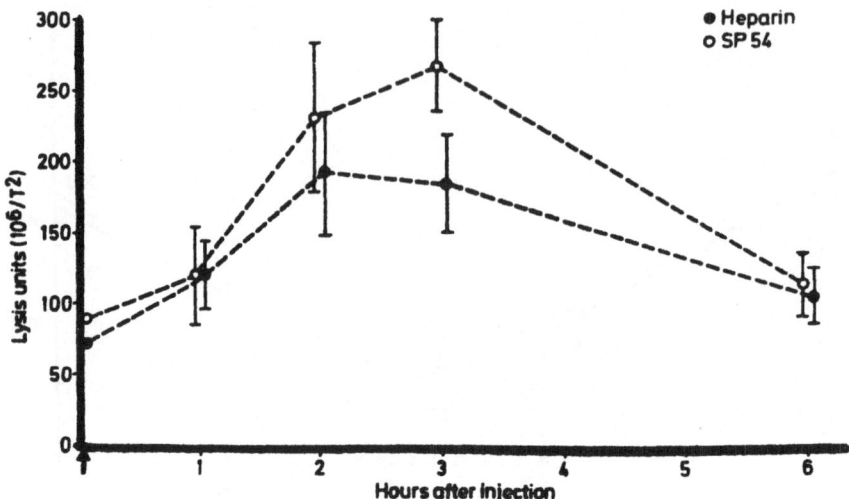

Fig. 6. As for Fig. 1, measuring euglobulin clot lysis,
expressed in units (10^6 divided by the square
of the lysis time in minutes).

Penthosane Polysulphate

The last of the minor fibrinolytics (non-thrombolytics) is
SP54, which we administered orally in doses of 75 mg per day. In
recent studies it was observed that penthosane polysulphate activates
lysis inducing secretion of t-PA by the vascular walls and by some
blood cells (Gaffney et al., 1982). It activates lysis even if
administered subcutaneously (Fischer et al., 1982; Fig. 6).

Urokinase

The only fibrinolytic drug that belongs to the class of thrombolytics that we currently use is UK, since for the time being we
have not yet received t-PA. Purified UK has never created any major
complications in our long experience. Since it is of human origin
it does not create any allergic reactions or fevers. As it induces
average fibrinolytic activation, it does not cause any significant
hemorrhages. In fact, with UK plasminogen depletion is never
extreme; the same applies to fibrinogen which rarely falls below
70 mg; the thrombin time does not rise excessively, and finally
the $alpha_2$-antiplasmin is never completely depleted.

It is for this reason that with UK hematic activation of the
lysis for dissolving the thrombus counts for little. UK induces

Table 6.   Angiographic Data during Thrombolytic Therapy
with UK in Acute Arterial Obliteration°

Cases Treated	Total Lysis	Partial Lysis	Unchanged cases
46	29 (63%)	11 (23%)	6 (14%)

°Obliterations were generally thrombotic. Thrombus age
was within 1 month. Obliterated arteries consisted of
the iliac, femoral, and popliteal.

endothrombolysis not esothrombolysis. The molecules that penetrate
the thrombus and activate the plasminogen to plasmin (the enzyme
that is active on fibrin) are useful for thrombolysis from UK.
The plasmin generated inside the thrombus can lyse the fibrin
because it is protected by the alpha$_2$-antiplasmin which does not
penetrate the thrombus. That part of UK which activates blood
plasminogen is not utilized for thrombolysis, because with UK there
is no depletion of the alpha$_2$-antiplasmin. The plasmin formed in
the blood is inactivated by the alpha$_2$-antiplasmin, since the
enzyme and the antienzyme unite in an irreversible bond.

The drug is mainly indicated for acute thrombotic arteriopathy
and deep venous thrombosis. In arteriopathy with acute thrombosis,
iliac, femoral, popliteal, UK is administered as follows: loading
dose - 500.000 I.U./15-20 minutes; maintenance dose - 300.000 I.U./
h/72 H (3 days). The results are satisfactory. The angiographic
data shows 63% total lysis and 23% partial lysis (Table 6).

In deep venous thrombosis, thrombolytic treatment must be
administered at the onset of the disease and not after the 7th day.
It must be remembered that often the onset is doubtful. It must
be maintained for a sufficient number of days to lyse the thrombus,
at least 6-7. It must be conducted with doses of UK that are
proportional to the extent of the thrombus, in general approximately
25-30 million units. Our therapeutic scheme in deep venous thrombosis is the following: loading dose - 500.000 I.U./15-20 minutes;
maintenance dose - 200.000 I.U./h/144 h (6 days).

The results, as seen from angiographic data, were as good as
early treatment. There were high percentages of total disobliter-

Table 7.   Phlebographic Data during Thrombolytic Therapy
           with UK in Deep Venous Thrombosis

Cases Treated	Thrombus Age	Total Lysis	Partial Lysis	Unchanged Cases
22	within 6 days	14 (69%)	6 (27%)	2 (4%)
12	within 1 month	4 (33%)	8 (67%)	–
8	within 2 months	–	8 (100%)	–
10	within 6 months	–	8 (80%)	2 (20%)

ation in those cases treated within the first 6 days of the onset.
In cases treated later, the number of total lysis decreased, even
if there was a satisfactory percentage of partial lysis (Table 7).

We shall not go into the merits of thrombolytic therapy with
SK, since the results are known in the literature, and mainly
because treatment with this drug has been abandoned by our group
for years. This was due to both the allergic and febrile reactions,
as well as hemorrhages that were often massive and sometimes fatal,
as a result of administration of the drug.

We want to emphasize that t-PA represents an important milestone
in thrombolytic therapy. It is highly possible that it will offer
new approaches in thrombolytic therapy with respect to classical
treatment. In vascular pathology, fibrin is also present in other
conditions with respect to acute obliterating arteriopathy and
deep venous thrombosis, and its removal could be a great step
forward. If we think that tissutal plasminogen activator is inactive
on blood proteins and takes as its own the fibrin receptors, it is
easy to believe that this drug could have rapid and simple lysis,
free of complications, and nearly ideal.

CONCLUSION

We do not want to end this chapter with simplifications or undue
optimism, even if a great deal of progress has been made with respect
to 10 years ago. The open problems in the treatment of angiologic
diseases are many, and the desire to improve our capabilities is

ever present. There is one concept that must be clearly expressed. In our opinion, we will never reach a conclusion or solution if we only use one, no matter how important, drug in this field. Atherosclerotic disease and thrombosis in general, whether arterial or venous, are multifactorial events. We are directing our attempts to changing the natural evolution of the most important angiologic diseases with treatments that are articulated in various directions and carried out with important drugs, which could lead not to solutions, but to major progress.

REFERENCES

Andreotti, L., and Nuzzaci, G., 1961, Modificazioni della fibrinolisi indotte dalla somministrazione endoarteriosa di acetilcolina, Riv. Crit. Clin. Med., 61:115.

Davidson, J.F., Walker, I.D., and Macallum, H.I., 1975, Study on the mechanism of action of anabolic steroids on fibrinolysis, in: "Progress in Chemical Fibrinolysis and Thrombolysis," Raven Press, New York.

Fischer, A.M., Merton, R.E., Marsch, N.A., Williams, S., Gaffney, P.G., Barrowcliffe, T.W., and Thomas, D.P., 1982, A comparison of penthosane sulphate and heparin. II. Effect of subcutaneous injection, Thromb. Haemost., 47:109.

Gaffney, P.G., Marsch, N.A., and Thomas, D.P., 1982, The influence of heparin and heparin-like substances on the fibrinolytic system in vivo, Seventh International Congress on Thrombosis, Valencia, October 13-16, Abstract, Basel.

Pratesi, F., Andreotti, L., Nuzzaci, G., Panerai, A., and Frosecchi, M., 1962, L'acetilcolina a dosi massive nelle arteriopatie obliteranti, Folia Angiologica, 9:36.

Tesi, M., 1975, Aspects of clinical methodology in fibrinolytic therapy, in: "Progress in Chemical Fibrinolysis and Thrombolysis," Raven Press, New York.

Tesi, M., Carini, A., Torrini, F., Trechas, P., and Bronchi, G.F., 1980, Il trattamento col mesoglicano solfato nella terapia dell'arteriopatia aterosclerotica, Riv. Farmacologia e Terapia, XI:309.

Walker, I.D., and Davidson, J.F., 1978, Long-term fibrinolytic enhancement with anabolic steroid therapy, in: "Progress in Chemical Fibrinolysis and Thrombolysis," Raven Press, New York.

THROMBOLYTIC PROPERTIES OF HUMAN

TISSUE-TYPE PLASMINOGEN ACTIVATOR

Désiré Collen

Center for Thrombosis & Vascular Research
Dept. of Medical Research, K.U. Leuven
Campus Gasthuisberg, Herestraat 49
3000 Leuven, Belgium

Human tissue-type plasminogen activator, isolated from the culture fluid of human melanoma cells, is a single chain protein with a molecular weight of approximately 72,000, which is indistinguishable from the plasminogen activator isolated from human uterus on the basis of molecular weight, amino acid composition, kinetic properties towards synthetic substrates, and immunochemical properties (Rijken and Collen, 1981). It is, however, easily converted by plasmin to a two-chain protein (Walléen et al., 1981; Rijken and Collen, 1981), with similar kinetic properties towards plasminogen (Rijken et al., 1982), but not towards low molecular weight synthetic substrates or inhibitors (Korninger and Collen, 1981; Ranby et al., 1982). It may be obtained in centigram quantities from melanoma cell culture fluid (Collen et al., 1982) which has allowed us to perform a series of studies, not only on the biochemical and biological properties of this activator, but also on its clinical value as a thrombolytic agent.

The thrombolytic effect of tissue-type plasminogen activator (two-chain form) was investigated in rabbits with an experimental pulmonary embolus (Matsuo et al., 1981). From this study it appeared that on a molar basis tissue-type plasminogen activator has a higher specific thrombolytic effect than urokinase, and that (partial) thrombolysis can be obtained without systemic fibrinolytic activation and breakdown of the hemostatic system.

The thrombolytic effect of tissue-type plasminogen activator was also investigated in dogs with an experimental thrombosis of the femoral vein (Korninger et al., 1982). Urokinase infusion at a rate of 2,500 IU per Kg per hour for four hours did not induce significant lysis (17.4 ± 3.7%) as compared to saline infusion (16.3 ± 3.8%). Significant lysis was obtained with 25,000 IU of urokinase per Kg per hour for four hours (40.6 ± 4.8%), but this was associated with defibrinogenation. Infusion of 2,500 urokinase equivalent units of plasminogen activator per Kg per hour for four hours caused significant lysis (33.5 ± 7.8% with single chain activator and 60.1 ± 10.8% with two chain activator), without causing any fibrinogen breakdown.

The influence of the molecular form of tissue-type plasminogen activator, the age of the thrombus, and the route of administration on the lysability of a thrombus were investigated in rabbits with experimental jugular vein thrombosis (Collen et al., 1983). One-chain and two-chain tissue-type plasminogen activator had very similar thrombolytic properties. Systemic infusion resulted in a dose-dependent degree of thrombolysis. The activator-induced thrombolysis, following infusion of 100,000 IU of tissue-type plasminogen activator (approximately 1 mg), was about 75% for fresh clots, 35% for 1 day old clots, 30% for 3 day old clots, and 50% for 7 day old clots. The thrombolytic activity of urokinase was more than 5 times lower than that of tissue-type plasminogen activator: infusion of 500,000 IU resulted in approximately 40% lysis of fresh clots and 25% of 1-3 day old clots, while 7 day old clots appeared to have become resistent to urokinase. Local infusion resulted in a 5 to 10 times higher thrombolytic effect of both tissue-type plasminogen activator and urokinase. Thrombolysis with tissue-type plasminogen activator was not associated with systemic activation of the fibrinolytic system as evidenced by unaltered plasma levels of fibrinogen, plasminogen, and 2-anti-plasmin. Systemic infusion of urokinase resulted in significant thrombolysis only at doses which were associated with systemic plasminogen activation. Local infusion of urokinase required a 5- to 10-fold higher dose than tissue-type plasminogen activator to obtain a similar degree of thrombolysis, which also occurred in the absence of systemic activation of the fibrinolytic system. From this study, it is concluded that the extent of thrombolysis by tissue-type plasminogen activator is mainly determined by the dose of activator, and its delivery to the thrombus, and much less by the age of the thrombus, or the molecular form of the activator.

The potential value of tissue-type plasminogen activator for
the treatment of acute myocardial infarction was established in an
experimental closed chest model in dogs (Bergmann et al., 1983).
Coronary thrombolysis in dogs was accomplished by intravenous infusion of tissue-type plasminogen activator at a rate of 10,000 IU
per min. within 10 min. Coronary recanalization was obtained
without inducing a systemic fibrinolytic state. Partial restoration of intermediary metabolism and nutritional myocardial blood
flow detected by positron emission tomography was achieved. It
was concluded that tissue-type plasminogen activator offers promise
for widely applicable, prompt, safe dissolution of coronary thrombi
accompanied by restitution of metabolism in jeopardized myocardium
in patients, without the need for cardiac catheterization and the
unavoidable attendant morbidity, risk, delay, and cost.

So far the thrombolytic properties of tissue-type plasminogen
activator were studied in two patients with ileofemoral vein
thrombosis (Weimar et al., 1981). Intravenous infusion of 5-10 mg
of tissue-type plasminogen activator over 24 hours resulted in
clearance of the vein in both patients. Thrombolysis was not associated with systemic activation of the fibrinolytic system
since no changes in fibrinogen, plasminogen, or 2-plasmin levels
were observed.

These studies indicate that tissue-type plasminogen activator
is a better thrombolytic agent than urokinase both in terms of
specific activity and in the absence of systemic fibrinolytic activation. Production of this material from tissue culture fluids is,
however, prohibitively laborious and expensive for large scale
studies, and for routine clinical application.

Recently, tissue-type plasminogen activator was produced by
recombinant DNA technology (Pennica et al., 1983). Studies aiming
to establish whether this substance might constitute an alternative
and more practical source of tissue-type plasminogen activator have
recently been initiated.

A comparison of recombinant tissue-type plasminogen activator
with that of melanoma origin revealed that both types had very
similar thrombolytic properties in rabbits with experimental jugular
vein thrombosis (Collen et al., unpublished). Thus, intravenous
infusion of 50,000 IU per Kg body weight of either substance over

4 hours yielded approximately 50% thrombolysis in the absence of systemic fibrinolytic activation in the blood.

In conclusion, it appears at present that tissue-type plasminogen activator might constitute a more specific and more selective thrombolytic agent than those presently available. Its clinical value remains, however, to be established.

REFERENCES

Bergmann, S.R., Fox, K.A.A., Ter-Pogossian, M.M., Sobel, B.E., and Collen, D., 1983, Clot-selective coronary thrombolysis, Science, (in press).

Collen, D., Rijken, D.D., Van Damme, J., and Billiau, A., 1982, Purification of human tissue-type plasminogen activator in centigram quantities from human melanoma cell culture fluid and its conditioning for use in vivo, Thromb. Haemost., 48:294.

Collen, D., Stassen, J.M., and Verstraete, M., 1983, Thrombolysis with human extrinsic (tissue-type) plasminogen activator in rabbits with experimental jugular vein thrombosis. Effect of molecular form and dose of activator, age of the thrombus and route of administration, J. Clin. Invest., 71:368.

Korninger, C., and Collen, D., 1981, Neutralization of human extrinsic (tissue-type) plasminogen activator in human plasma: no evidence for a specific inhibitor, Thromb. Haemost., 46:662.

Korninger, C., Matsuo, O., Suy, R., Stasse, J.M., and Collen, D., 1982, Thrombolysis with human extrinsic (tissue-type) plasminogen activator in dogs femoral vein thrombosis, J. Clin. Invest., 69:573.

Matsuo, O., Rijken, D.C., and Collen, D., 1981, Thrombolysis by human tissue plasminogen activator and urokinase in rabbits with experimental pulmonary embolus, Nature, 291:590.

Pennica, D., Holmes, W.E., Kohr, W.J., Harkins, R.N., Vehar, G.A., Ward, C.A., Bennett, W.F., Yelverton, E., Seeburg, P.H., Heyneker, H.L., Goeddel, D.V., and Collen, D., 1983, Cloning and expression of human tissue-type plasminogen activator cDNA in E.coli, Nature, 301:214.

Ranby, M., Bergsdorf, N., and Nilsson, T., 1982, Enzymatic properties of the one- and two-chain form of tissue plasminogen activator, Thromb. Res., (in press).

Rijken, D.C., and Collen, D., 1981, Purification and characterization of the plasminogen activator secreted by human melanoma cells in culture, J. Biol. Chem., 256:7035.

Rijken, D.C., Hoylaerts, M., and Collen D., 1982, Fibrinolytic properties of one-chain and two-chain human extrinsic (tissue-type) plasminogen activator, J. Biol. Chem., 257:2920.

Walléen, P., Ranby, M., Bergsdorf, N., and Kok, P., 1981, Purification and characterization of tissue plasminogen activator: on the occurrence of two different forms and their enzymatic properties, in: "Progress in Fibrinolysis," Vol. V, J.F. Davidson, I.M. Nilsson, and B. Astedt, eds., Churchill Livingstone, Edinburgh.

Weimar, W., Stibbe, J., van Seyen, A.J., Billiau, A., De Somer, P., and Collen, D., 1981, Specific lysis of an iliofemoral thrombus by administration of extrinsic (tissue-type) plasminogen activator, Lancet, ii:1018.

THE ANTICOAGULANT PROPERTIES OF

FRACTIONATED AND DEPOLYMERIZED HEPARINS

Jeffrey Denton and
David A. Lane

Department of Hematology
Charing Cross Hospital Medical School
Hammersmith, London W6 8RF

INTRODUCTION

The anticoagulant property of heparin was first recognized by McLean (1916) during an investigation of the properties of chloroform extracts of various tissues. The name heparin was given to this material because of its hepatic origins. Plasma cofactor was shown to be necessary for its activity by Howell and Holt (1918) and purified heparin polysaccharide was first isolated by Howell (1925). The development of heparin for clinical research took place over forty years ago by research groups in Toronto and Stockholm (Jorpes 1939; Murray et al., 1937) and has now become universally used as an essential anticoagulant during surgery and enjoys widespread use in the control and prevention of thromboembolism postoperatively.

From a clinical viewpoint commercial heparin can be regarded as similar material today as when it was first marketed. Heparin preparations do, however, vary from manufacturer to manufacturer and from batch to batch (Lane et al., 1978). Chemically heparin is an alternating copolymer of uronic acid and glucosamine. The uronic acid is either glucuronic acid or its C5 epimer iduronic acid. Iduronic acid usually has a sulphate ester group at position 2, while the glucosamine can have N-sulphate, N-acetyl or free amino groups and a 6-sulphate ester group. The molecular weight of heparin chains varies from 5,000 to 30,000. Consequently, commercial heparin preparations are highly polydisperse and require suitable separation methods to prepare defined fractions before detailed studies of its anticoagulant properties can be made.

FRACTIONATION METHODS

Early methods of heparin fractionation by alcohol solubility and quaternary ammonium salt precipitation were designed to isolate heparin from the other glycosaminoglycans rather than for the purpose of examination of its molecular properties. However, Lasker and Stivala (1966) showed that heparin's anticoagulant activity increased with the increasing molecular weight of the polysaccharide.

Gel filtration has been a commonly used method of examining the molecular weight of heparin and the technique has been used to study its molecular weight distribution and to prepare heparin fractions of defined molecular weight. Experiments carried out by Cifonelli (1974) and Johnson and Mulloy (1976) confirmed that the anticoagulant activity increased with increasing molecular weight over most of the molecular weight range. In these studies anticoagulant activities were measured by pharmacopeial or clotting assays. The development of assays which measure the ability of heparin to potentiate the inhibition of factor Xa by Yin et al. (1973) and Denson and Bonner (1974) was an important step in studying the anticoagulant behavior of heparin. Anderson et al. (1976) prepared heparin fractions by gel filtration and assayed them by an activated partial thromboplastin time (APTT) assay and by an anti-factor Xa assay. The expected increase in potency with increasing molecular weight in APTT assays was observed, but the highest anti-factor Xa activity was located in the lowest molecular weight fraction. It has been since shown (MacGregor et al., 1979 and 1980) that the high anti-factor Xa potency of low molecular weight heparin prepared by gel filtration can be in part explained by the antiheparin activity of low density lipoprotein which selectively inhibits the anti-factor Xa activity of high molecular weight heparin. Any remaining high anti-factor Xa activity is probably due to the structural requirements of the heparin-inhibitor-proteinase interactions (see below).

Perhaps the most important method of heparin fractionation with regard to understanding its anticoagulant activity is affinity chromatography. As stated earlier, heparin requires a plasma protein cofactor which has been identified as the α2 glycoprotein antithrombin, AT III (Abildgaard, 1968; Rosenberg and Damus, 1973). Experiments performed by Höök et al. (1976) and by Lam et al. (1976) showed that heparin could be separated into two or three distinct fractions with respect to its binding to AT III. Fractions which did not bind to the protein, which bound weakly and which bound strongly were prepared and were designated no-affinity (NA), low affinity (LA), and high affinity (HA) heparin respectively. It was found that the anticoagulant activity of the parent heparin was almost exclusively confined to the HA heparin fraction (LA heparin was 19 u/mg, HA heparin 285 u/mg), which contained 30% of the total polysaccharide material. These experiments implied that HA heparin has unique structural characteristics that enable it to bind to AT III and determine its anticoagulant activity.

DEPOLYMERIZATION METHODS

A method of studying the structure of heparin which is complementary to fractionation methods discussed above is depolymerization of the heparin chain. There are three ways by which this can be achieved: (i) nitrous acid depolymerization (Lagunoff and Warren, 1962; Cifonelli and King, 1972; Shirley and Conrad, 1976), (ii) enzymatic cleavage (Lindahl et al., 1979), and (iii) periodate oxidation followed by alkaline hydrolysis (Fransson, 1978). Treatment of heparin with nitrous acid brings about reductive cleavage of the polysaccharide chains after N-sulphated and unsubstituted hexosamines, and if allowed to proceed to completion would produce a mixture of disaccharides and tetrasaccharides with 2.5-anhydromanhose groups at the reducing end. Enzymatic depolymerization of heparin is carried out with the enzyme from the bacterium Flavobacterium heparinum. The enzyme is a β-eliminase which cleaves the heparin molecule after N-sulphated glucosamine groups. Periodate oxidizes the 2 - 3 diol groups of glucuronic acid or non-sulphated iduronic acid to aldehyde moieties. Selective oxidation of glucuronic acid can be achieved by reducing the pH and temperature, while both uronic acids are oxidized at pH 7 and 37°C (Fransson, 1978; Fransson et al., 1980). Subsequent treatment of the oxidized heparin with alkali released the polysaccharide from the reducing end of the oxidized uronic acid.

By controlling the extent of these reactions it is possible to produce heparin chains of specific size. Hopwood et al. (1976)

showed that if heparin bound to immobilized AT III was digested
with heparinase, the heparin fragment that remained bount to AT III
was a tetradecasaccharide. Lindahl et al. (1979) showed that similar sized heparin fragments could be prepared by partially de-
polymerizing heparin with heparinase or nitrous acid, followed by
affinity chromatography on AT III - Sepharose. The material prepared
in this way had very striking anticoagulant properties (Thunberg et
al., 1979). Potencies measured by pharmacopeial, APTT or thrombin-
specific assays were minimal, yet an anti-factor Xa specific activity of 1,800 u/mg was attained. The majority of this very high
anti-factor Xa activity of heparin oligosaccharides arises from the
presence of structural features, present in the parent molecules
and retained after depolymerization, that enable heparin to bind
with high affinity to AT III: this region of the heparin molecule
is termed the AT III binding site.

STRUCTURE OF THE AT III - BINDING SITE OF HEPARIN

If heparin is nearly exhaustively degraded by nitrous acid
treatment, then subjected to affinity chromatography on immobilized
AT III, and material which binds to the column at the ionic strength
of 1 M NaCl is recovered, it can be shown by Sephadex G50 gel filtration that the smallest fragment which can bind to the protein is an
octasaccharide. The structure of the AT III binding octasaccharide
has been determined (Thunberg et al., 1980). The trisulphated
glucosamine residue located at position 4 of the octasaccharide
contains a 3-O-sulphate ester group unique to the AT III binding
fragment. If the octasaccharide is de-N-sulphated, AT III binding
is destroyed (Riesenfeld et al., 1981) but the removal of the
iduronic acid at the non-reducing end of the fragment also does not
affect AT III binding (Thunberg et al., 1982). If the hexasaccharide
(1 - 6) is prepared by nitrous acid treatment of the octasaccharide
followed by Sephadex G50 gel filtration, its AT III binding affinity
is reduced tenfold because of the loss of the N-sulphate group at
position 6 (Thunberg et al., 1982). Similarly, removal of the
N-acetyl-glucosamine 6-sulphate group by breaking the chain after
residue 2 by hydrazinolysis and nitrous acid treatment at pH 4.0
destroys affinity for AT III (Thunberg et al., 1982). The above
chemical modification studies and NMR studies (Casu et al., 1981)
have indicated that the pentasaccharide comprising residues 2 - 6
of the octasaccharide contains all the structural requirements for
AT III binding.

MECHANISM OF ACTIVITY AS A PROTEINASE INHIBITOR

The anticoagulant activity of heparin is intimately linked
with that of the inhibitor AT III. When heparin binds to AT III
it induces a conformational change in the protein (Rosenberg and
Damus, 1973) which can be detected by fluorescence (Nordenman et
al., 1978). This conformational change makes the arginine active
site of the AT III more available, increasing the rate at which
proteinase-inhibitor complex is formed. It was originally postulated
that heparin could accelerate the inhibiting effect of AT III
against all the coagulation system enzymes, but this has recently
been re-examined and questioned (Scott et al., 1982). A study of
the anticoagulant behavior of HA heparin oligosaccharides (prepared
by nitrous depolymerization) and of heparin fractions (prepared by
gel filtration) by Holmer et al. (1981) showed that the proteolytic
enzymes of the coagulation cascade could be classified into two
groups. Inhibition of thrombin, factors IXa and XIa showed similarities in their dependence on the molecular weight of heparin,
and the inhibitory potency of heparin was found to decrease with
decreasing molecular weight. Inactivation of factor Xa, factor
XIIa and kallikrein was less dependent on molecular weight, and an
oligosaccharide preparation which has no inhibitory activity on
the thrombin-like group of enzymes was found to have considerable
inhibitory activity against these enzymes. Jordan et al. (1980a,b)
have shown that the kinetics of inhibition of thrombin and factor
IXa are dependent on the heparin concentration. The rate of inhibition reaches a maximum when HA heparins and AT III are present
in equimolar amounts and then fall as the heparin concentration is
increased. The inhibition of factor Xa and plasmin also reaches
a maximum rate when HA heparin and AT III are equimolar, but this
does not change with increasing heparin concentration. Thrombin
and factor IXa bind to heparin strongly, whilst plasmin and factor
Xa did not bind with high avidity.

The mechanistic rationalization of the above results (Oosta
et al., 1980) is that heparin-activated AT III and factor Xa-like
enzymes combine directly to produce the enzyme inhibitor complex.
Inhibition of the thrombin-like enzymes requires that the enzyme
and AT III both must bind to heparin before enzyme-AT III complex
formation can be enhanced. Because both enzyme and inhibitor must
bind to the polysaccharide ('approximation' effect), a larger sugar
chain is required for this than just the AT III binding octasac-

charide. In the case of thrombin the minimum chain required for
activity has been reported to be either a hexadecasaccharide (Oosta
et al., 1980) or an octadecasaccharide (Denton et al., 1983a).

OTHER INTERACTIONS

Besides its interaction with AT III, two other proteins may
play a role in the expression of its anticoagulant activities,
platelet factor 4 (PF4) and heparin cofactor II. PF4 is a low
molecular weight platelet granule protein which is extruded during
the platelet release reaction. It is the most potent heparin
neutralizing agent in the blood and typically 1 mg of PF4 will
totally inhibit 1 mg of heparin. However, heparin oligosaccharides
with high anti-factor Xa activities are not readily inhibited
(Denton et al., 1983b) unless the material is at least a tetradecasaccharide (Denton et al., 1983a).

Heparin cofactor II is a newly discovered (Tollefsen and Blank,
1981) plasma protein which acts as a coagulation inhibitor in purified form, its action is accelerated by heparin and it is chemically
distinct from AT III. It is not known whether heparin cofactor II
has the same specific structural requirements for heparin binding
as AT III or what its physiological function is.

ANTICOAGULANT AND ANTITHROMBOTIC FUNCTION

It was suggested by Wessler and Yin (1973) that the prophylactic
effect of low doses of heparin may be achieved by its acceleration
of inhibition of factor Xa by AT III. Because of the central role
of factor Xa in the coagulation system, being activated by components
of the intrinsic and extrinsic pathways, its inhibition was thought
to play an important role in heparins anti-thrombotic activity.
The suggestion emerging from Wessler and Yin's work was that heparins
with high anti-factor Xa activity but lower effects on overall clotting (as measured by APTT-like assays) may be good antithrombotic
agents, yet have less hemorrhagic side effect than standard heparins.
Much recent work has been directed towards producing low molecular
weight heparins which have a high anti-factor Xa but lower APTT
(and antithrombin) activities. Suitable low molecular weight
heparins have been prepared by alcohol fractionation, gel filtration
and by controlled nitrous acid depolymerization. Animal model experiments have been carried out by Carter et al. (1982), in which

a low molecular weight (4,500) heparin prepared by nitrous acid
depolymerization and a standard heparin (molecular weight 15,500)
were given at equivalent anti-factor Xa doses to rabbits. The low
molecular weight material was less hemorrhagic, was less able to
inhibit collagen-induced platelet aggregation and, most importantly,
showed greater antithrombotic activity than the standard heparin.
This has provided the first experimental evidence that the antithrombotic and anticoagulant activities of heparin are not necessarily identical, and the low molecular weight heparin may be a
more suitable antithrombotic agent than unfractionated commercial
heparin. The results of a clinical trial designed to test the
ability of low molecular weight heparin to prevent postoperative
deep vein thrombosis have provided further support for this concept
(Kakkar et al., 1982).

REFERENCES

Abildgaard, U., 1968, Highly purified antithrombin III with haparin
    cofactor activity prepared by disc electrophoresis, Scand.
    J. Lab. Invest., 21:89-91.
Anderson, L.O., Barrowcliffe, T.W., Holmer, E., Johnson, E.A., and
    Sims, G.E.C., 1976, Anticoagulant properties of heparin
    fractionated by affinity chromatography and gel filtration,
    Thromb. Res., 9:575-583.
Carter, C.J., Kelton, J.G., Hirsh, J., Cerskus, A., Santos, A.V.,
    and Gent, M., 1982, The relationship between the hemorrhagic
    and antithrombotic properties of low molecular weight heparin
    in rabbits, Blood, 59:1239-1245.
Casu, B., Oreste, P., Torri, G., Zoppett, G., Choay, J., Lormeau,
    J.C., Petitou, M., and Sinaÿ, P., 1981, The structure of
    heparin oligosaccharide fragments with high anti-factor Xa
    activity containing the minimal antithrombin III binding
    site, Biochem. J., 197:599-609.
Cifonelli, J.A., and King, J., 1972, The distribution of 2-acetamido-
    2-deoxy-D-glucose residues in mammalian heparins, Carbohyd.
    Res., 21:173-186.
Cifonelli, J.A., 1974, The relationship of molecular weight and
    sulphate content and distribution to anticoagulant activity
    of heparin preparations, Carbohyd. Res., 37:147-154.
Denson, K.W.E., and Bonner, J., 1973, The measurement of heparin.
    A method based on the potentiation of anti-factor Xa, Thromb.
    Diath. Haem., 30:471-482.

Denton, J., Lane, D.A., Slater, A.M., Thunberg, L., and Lindahl, U.,
    1983, Size of heparin oligosaccharides required for anti-
    thrombin activity and interaction with platelet factor 4
    (PF4). Submitted to IXth International Congress on thrombosis
    and hemostasis.

Denton, J., Lane, D.A., Slater, A.M., Thunberg, L., and Lindahl, U.,
    1983, Binding of platelet factor 4 to heparin oligosaccha-
    rides, Biochem. J., 209:455-460.

Fransson, L.A., 1978, Periodate oxidation of the D-glucuronic acid
    residues in heparin sulphate and heparin, Carbohyd. Res.,
    62:235-244.

Fransson, L.A., Malmström, A., Syöberg, I., and Huckerby, T.N.,
    1980, Periodate oxidation and alkaline degradation of heparin
    related glycans, Carbohyd. Res., 80:131-145.

Holmer, E., Kurachi, K., and Söderström, G., 1981, The molecular
    weight dependence of the rate enhancing effect of heparin
    on the inhibition of thrombin, factor Xa, factor IXa, factor
    XIa, factor XIIa and kallikrein by antithrombin, Biochem. J.,
    193:395-400.

Höök, M., Björk, I., Hopwood, J., and Lindahl, U., 1976, Anticoagu-
    lant activity of heparin: separation of high-activity species
    by affinity chromatography on immobilized antithrombin,
    F.E.B.S. Lett., 66:90-93.

Hopwood, J., Höök, M., Linker, A., and Lindahl, U., 1976, Anticoagu-
    lant activity of heparin: isolation of antithrombin binding
    sites, F.E.B.S. Lett., 69:51-54.

Howell, W.H., and Holt, E.O., 1918, Two new factors in blood coagu-
    lation - heparin and proantithrombin, Amer. J. Physiol., 47:
    328-341.

Howell, W.H., 1925, The purification of heparin and its presence in
    blood, Amer. J. Physiol., 71:553-562.

Johnson, G.A., and Mulloy, B., 1976, The molecular weight range of
    mucosal heparin preparations, Carbohyd. Res., 51:119-127.

Jordan, R.E., Oosta, G.M., Gardner, W.T., and Rosenberg, R.D., 1980,
    The binding of low molecular weight heparin to hemostatic
    enzymes, J. Biol. Chem., 255:10073-10080.

Jordan, R.E., Oosta, G.M., Gardner, W.T., and Rosenberg, R.D., 1980,
    The kinetics of hemostatic enzyme-antithrombin interactions
    in the presence of low molecular weight heparin, J. Biol.
    Chem., 225:10081-10090.

Jorpes, E., 1939, "Heparin: Its Chemistry, Physiology and Applica-
    tion in Medicine," Oxford University Press, London.

Kakkar, V.V., Djazaeri, B., Fok, J., Fletcher, M., Scully, M.F., and Westwick, J., 1982, Low molecular weight heparin and prevention of deep vein thrombosis, Brit. Med. J., 284: 375-379.

Lagnuff, D., and Warren, G., 1962, Determination of 2-deoxy-2-sulphaminohexose content of mucopolysaccharides, Arch. Biochem. Biophys., 99:396-400.

Lam, L.H., Silbert, J.E., and Rosenberg, R.D., 1976, Separation of active and inactive forms of heparin, Biochem. Biophys. Res. Commun., 69:570-576.

Lane, D.A., MacGregor, I.R., Michalski, R., and Kakkar, V.V., 1978, Anticoagulant activities of four unfractionated and fractionated heparins, Thromb. Res., 12:257-271.

Lasker, S.E., and Stivala, S.S., 1966, Physiochemical studies of fractionated bovine heparin I. Some dilute solution properties, Arch. Biochem. Biophys., 115:360-372.

Laurent, T.C., 1961, Studies on fractionated heparin, Arch. Biochem. Biophys., 92:224-231.

Lindahl, U., Bäckström, G., Höök, M., Thunberg, L., Fransson, L.A., and Linker, A., 1979, Structure of the antithrombin binding site in heparin, Proc. Natl. Acad. Sci. U.S.A., 76:3198-3202.

Lindahl, U., Bäckström, G., Thunberg, L., and Leder, I.G., 1980, Evidence for a 3-0 sulphated D-glucosamine residue in the antithrombin binding sequence of heparin, Proc. Natl. Acad. Sci. U.S.A., 77:6551-6555.

MacGregor, I.R., Lane, D.A., and Kakkar, V.V., 1979, Evidence for a plasma inhibitor of the heparin accelerated inhibition of factor Xa by antithrombin III, Biochem. Biphys. Acta, 586: 584-593.

MacGregor, I.R., Lane, D.A., and Kakkar, V.V., 1980, The anti heparin properties of human low density lipoprotein, Biochem. Biophys. Acta, 617:472-479.

McLean, J., 1916, The thromboplastic action of cephalin, Amer. J. Physiol., 25:683-687.

Murray, D.W.G., Jaques, L.B., Perrett, T.S., and Best, C.G., 1937, Heparin and the thrombosis of veins following injury, Surgery, 2:163-187.

Nordenman, B., Danielsson, A., and Björk, I., 1978, The binding of low affinity and high affinity heparin to antithrombin, Eur. J. Biochem., 90:1-6.

Nordenman, B., and Björk, I., 1978, Binding of low affinity and high affinity heparin to antithrombin. Ultraviolet spectroscopy and circular dichroism studies, Biochemistry, 17:3339-3344.

Oosta, G.M., Gardner, W.T., Beeler, D.L., and Rosenberg, R.D., 1980, Multiple functional domains of the heparin molecule, Proc. Natl. Acad. Sci. U.S.A., 78:829-833.

Riesenfeld, J., Thunberg, L., Höök, M., and Lindahl, U., 1981, The antithrombin binding sequence of heparin: Location of essential N-sulphate group, J. Biol. Chem., 256:2389-2394.

Rosenberg, R.D., and Damus, P.S., 1973, The purification and mechanism of action of human antithrombin-heparin cofactor, J. Biol. Chem., 248:6490-6505.

Scott, C.F., Schapira, M., and Coleman, R.W., 1982, Effect of heparin on the inactivation rate of human factor XIa antithrombin III, Blood, 60:940-947.

Shirley, J.E., and Conrad, H.E., 1976, Formation of anhydrosugars in the chemical depolymerization of heparin, Biochemistry, 15:3932-3942.

Thunberg, L., Lindahl, U., Tengblad, A., Laurent, T.C., and Jackson, C.M., 1979, On the molecular-weight dependence of the anticoagulant activity of heparin, Biochem. J., 181:241-243.

Thunberg, L., Bäckström, G., Grundberg, H., Riesenfeld, J., and Lindahl, U., 1980, The molecular size of the antithrombin binding sequence in heparin, F.E.B.S. Lett., 117:203-205.

Thunberg, L., Bäckström, G., and Lindahl, U., 1982, Further characterization of the antithrombin-binding sequence in heparin, Carbohyd. Res., 100:393-410.

Tollefsen, D.M., and Blank, M.K., 1981, Detection of a new heparin dependent inhibitor of thrombin in human plasma, J. Clin. Invest., 68:589-596.

Wessler, S., and Yin, E.T., 1973, The theory and practice of minidose heparin in surgical patients: a status report, Circulation, 47:671-676.

Yin, E.T., Wessler, S., Butler, J.V., and Cole, S., 1973, Plasma heparin: a unique, practical, submicrogram-sensitive assay, J. Lab. Clin. Med., 81:298-310.

CLINICAL AND BIOLOGICAL ASPECTS OF

ESSENTIAL THROMBOCYTHEMIA

S. Tura, L. Gugliotta, M. Baccarani, C. Bernasconi,
C. Biagini, P. Boccaccio, A. Canevari, G. Cavazzini,
L. Chetti, A. Grossi, A. Guarini, G. Isacchi, F. Mandelli,
P.M. Mannucci, M. Morfini, F.I. Pareti, P. Ricci,
D. Russo, P.L. Rossi Ferrini, E. Salvidio, and A. Zaccaria

Essential Thrombocythemia Italian Study Group
Institute of Hematology
University of Bologna
Bologna, Italy

INTRODUCTION

Essential thrombocythemia (ET) is a clonal myeloproliferative disease that involves a multipotent stem cell common to the mega-karyocytic, granulocytic, and erythrocytic lines (16). In ET the major abnormality is represented by the increase of the megakaryocyte number, volume, and mass (13,25,26,45). The resulting platelet production is usually 4-6 times more than the normal value, while the platelet survival is normal or slightly decreased (25,29,34,43,45).

ET is the prototype of "autonomous" thrombocytosis where the platelet production is unresponsive to normal regulatory processes, and the platelet count does not correlate with the megakaryocyte volume (25,26,45). Thrombokinetics, clinical and laboratory features are usually adeguate to differentiate ET from the reactive thrombo-cytosis (23,43-46), and from the other chronic myeloproliferative diseases (polycythemia vera, myelofibrosis with myeloid metaplasia, chronic myeloid leukemia) (11,23,30,43,45,46).

In ET, frequently characterized by thrombotic and/ hemor-rhagic syndromes, the thrombocytosis is persistent and the platelet

Table 1. Clinical and Laboratory Data in 93 Patients with
          Essential Thrombocythemia at Diagnosis (ETIGS 1983)

Number of patients	93
Males/Females	39/54
Median age (range)	46 (8-78)
Thrombosis	31 pts (33%)
Hemorrhage	31 pts (33%)
Hepatomegaly	38 pts (41%)
Splenomegaly	50 pts (53%)
Platelets>1,000 x 10^9/1	62 pts (66%)
Hb<11 g/dl	23 pts (30%)
Leukocytes>9 x 10^9/1	68 pts (73%)

count is usually over 700-1,000 x 10^9/1 (23,43,46). In some cases,
ET is observed in patients splenectomized for congestive splenomegaly, splenic infarction, or splenic traumatic rupture (23,43).
In these patients, splenectomy appears to have unmasked previously
latent thrombocythemia (2,23,24,43,45).

CLINICAL AND HEMATOLOGICAL ASPECTS

Sex and Age

     ET, distributed more or less equally in men and women (23,43,
46), is most commonly observed in middle-aged and old-aged people,
although cases have been documented in younger people and even in
children (1,23,31,43,46). In the updated report of the Essential
Thrombocythemia Italian Study Group (ETIGS) in 93 patients (16 of
them were previously splenectomized), mean and median ages were
46 years (range 8-78). Seventeen out of 93 patients (18%) were
aged between 8 and 30 (Table 1). Familial occurrence of thrombocythemia has been reported in a few cases (1,14,43).

Thrombosis and Hemorrhage

     Thrombotic and/or hemorrhagic episodes are frequent in ET
(23,30). Thrombosis develops frequently in a splenic vein, the
superficial and deep veins of the legs, the pulmonary region, and
the digital vessels. It may also involve other sites such as the
hepatic veins and the penial veins. Gum bleeding, hematomas,

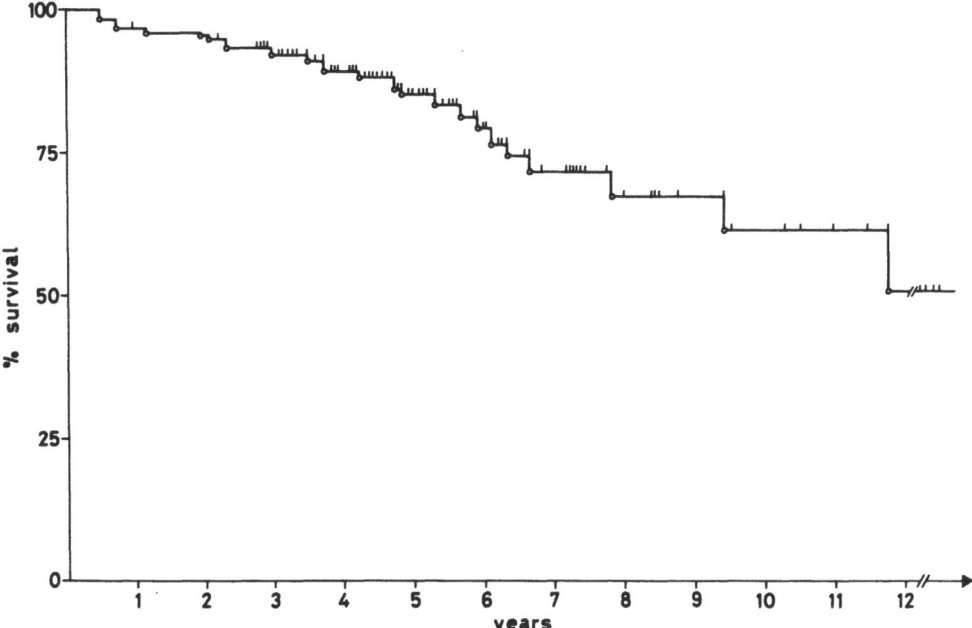

Fig. 1.   Survival of 93 patients with essential thrombocythemia
           (ETIGS 1983).

epistaxis, gastrointestinal bleeding, excessive bleeding after dental
extraction, and menorrhagia are the most common hemorrhagic
manifestations (23,30,43,46). Thrombosis and hemorrhage are
usually more frequent in the patients with severe thrombocytosis
(23,43,46). In the ETIGS report, thrombosis and hemorrhage were
observed in 31 (33%) and 31 (33%) patients respectively, with
higher incidence in the splenectomized patients (Table 1).

Splenomegaly and Hepatomegaly

     In the above mentioned report, splenomegaly and hepatomegaly
were observed in 50 (53%) and 38 (41%) patients respectively
(Table 1). The incidence of hepatomegaly and splenomegaly showed
itself to be independent from the platelet count (43).

Bone Marrow, Platelets, Erythrocytes, and Leukocytes

     A moderate hyperplasia of granulocytic and erythrocytic lines
was associated, in about half the patients, to a constant increase
of megakaryocyte mass (23,43,46). A bone biopsy showed, in

approximately 1/3 of the cases, a variable degree of fibrosis (43).
The platelet count was usually >700 x 10^9/1 and, frequently, it was
>1,000 x 10^9/1 (Table 1), particularly in splenectomized patients
(23,43,46). A thrombocytosis of lower entity may be one of the
causes of uncertainty in the diagnosis of ET (23,43,46).

Anemia (Hb <11 g/dl) was observed in 30% of the patients
(Table 1), and a moderate erythrocytosis was registered in 15-20%
of the patients (23,43). A leukocytosis (9-40 x 10^9/1) was
observed in more than 2/3 of the cases (Table 1), with neutrophilia,
slight eosinophilia, and basophylia; metamyeolocytes and occasional
myelocytes were found in peripheral blood of about 1/5 of the
patients (23,43).

Course and Evolution of the Disease

ET, if adequately controlled with cytotoxic agents, is a
disease which has a relatively benign course (6,15). In the
revised and updated report of the ETIGS (93 patients with mean
follow-up of 71 months), the median survival was about 12 years
(Fig. 1), and no differences were found between splenectomized
and non-splenectomized patients. Twenty-one out of 93 patients
died 6-141 months after the diagnosis: 10 of them died from
thrombosis, 6 from acute non-lymphoid leukemia, and only 2 from
hemorrhage (20).

BIOLOGICAL AND HEMOSTATIC ASPECTS

Various chromosomal abnormalities have been found at random
in about 1/3 of the patients. Deletion of a long arm of one
chromosome 21 (21q-), previously described (47), has not been
confirmed in a recent workshop.

In ET a prolonged bleeding time, observed in about 20% of
the cases, does not correlate with the severity of thrombocytosis,
or with the tendency to thrombohemorrhagic complications (7,38,43,
45). Bleeding and thrombosis are mainly related to the morpho-
functional abnormalities of the platelets (10,38,42,43,48). An
evident anisocytosis, with the presence of large and small platelets,
is well demonstrated by electron microscopy (1,3), and also by a
platelet size distribution curve (5,19,27). Platelet microcytosis
is prevalent and the Mean Platelet Volume (MPV) is lower than

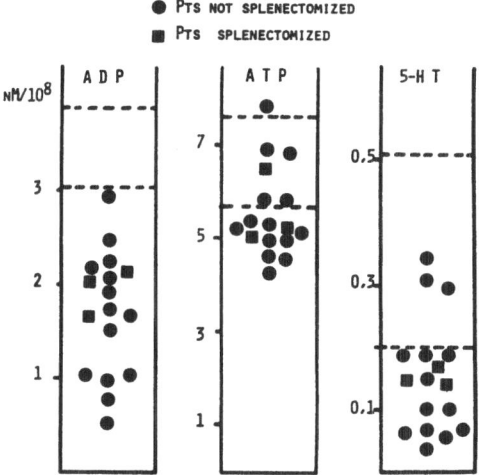

Fig. 2. Platelet levels of ADP,
ATP, and 5-HT in 17
patients with essential
thrombocythemia.

normal (5,19,27). A study in 30 patients showed an mPV of $82 + 0.6\mu^3$
(N.V. 9.3+0.8) (19). In more than 50% of the patients, the
platelet aggregation, and particularly the release reaction, is
defective when ADP, adrenaline, and collagen are used as inducers
(18,38,42,43,48).

This defect, evaluated by a photometric method, may also be
demonstrated by studying the morphology of the platelet aggregate,
with the use of electron microscopy (21). This defective platelet
aggregation is referred to an acquired Storage Pool Defect (SPD)
(1,8,38,40). In fact, the biochemical studies have demonstrated
an important reduction in ADP, ATP, and 5-HT stored in the dense
granules (Fig. 2) of platelets, and electron microscopy has shown
a severe decrease in the number of these granules (1,3,38,40).
This SPD seems to be related to the production of abnormal platelets
lacking in granules (6,35,38,40). In fact, SPD is present in ET
also when the platelet number is lowered to normal value, and it
is absent in patients with reactive thrombocytosis (38). The
metabolism of arachidonic acid may also be abnormal due to a
decrease of lipoxygenase and/or cyclo/oxygenase activity (36,40,41).
The platelet adhesiveness in ET is nearly always decreased (18,38,
43,45), and various biochemical abnormalities occur in the platelet
membrane: a deficiency of α-adrenergic receptors (28); a decreased

number of receptors for prostaglandin D_2 (9); an increase in the
number of intramembranous particles shown by the freeze-fracture
technique (3). In the platelet membrane, a decreased sialic acid
and an abnormal glycoprotein (GP) distribution are frequently
observed (4,12,22,32). GP I, GP II and GP III are usually decreased
while GP IV is normal or increased (4,12,22).

An interaction of these quantitative and qualitative platelet
abnormalities with cellular, plasmatic, vascular, and hemorheologic
factors is responsible for the appearance of thrombosis and/or
hemorrhage in ET (45,46).

CONCLUSION

The morpho-functional abnormality of the platelets (anysocyto-
sis, membrane defect, Storage Pool Defect) is characteristic, and
may be considered a sort of biological marker in ET and other
chronic myeloproliferative disorders (MPD).

Clinical, laboratory, and biological features are usually
adequate to differentiate the MPD between themselves (30,43,46).
For the diagnosis of ET, the ETIGS has pinpointed some necessary
parameters (platelet number $< 700 \times 10^9$/l for at least 6 months:
an increase of megakaryocytic mass in the absence of severe
myelofibrosis; and SPD in patients with normal red blood cell mass;
an absence of Ph^1 chromosome), and other complementary parameters
(thrombosis and/or hemorrhage; hepatosplenomegaly; leukocytosis
$< 40 \times 10^9$/l) (43). Now it is proposed to make use of this suggested biological marker of chronic MPD to make a diagnosis of ET
at the onset of the disease (without a 6-month period of observation), and also when thrombocytosis is $< 700 \times 10^9$/l. Being quite
sure about diagnosis allows us to treat the patients promptly
with cytotoxic agents (alkilating or antimitotic drugs or ^{32}phos-
phate), and a long-term thrombocytoapheresis may be avoided (17).
Moreover, the frequent ill-use of antiaggregating agents in
patients with complex platelet defect is avoidable (43,46).
Although ET may be considered a disease with a benign course
(median survival of 12 years), an appreciable number of patients
die from thrombotic complications probably due to an inadequate
use of cytotoxic drugs. In a few cases, ET evolves into acute
non-lymphoid leukemia, sometimes after a spell of myelofibrosis,
and it is difficult to say if the cytotoxic treatment plays a part
in this transformation.

REFERENCES

1. M.I. Barnhart, T.H. Kim, B.L. Evatt, A.H. Ragab, V.K. Lui,
    J. Herman, and J.M. Lusher, Essential thrombocythemia in
    a child: platelet ultrastructure and function, Am. J.
    Hematol., 8:87 (1980).
2. T.A. Bensinger, G.L. Logue, and R.W. Rundles, Hemorrhagic
    thrombocythemia: control of postsplenectomy thrombocytosis
    with melphanan, Blood, 36:61 (1970).
3. G. Biagini. L. Gugliotta, P. Preda, A. Guarini, G. Fumagalli,
    and R. Laschi, Platelets in primary thrombocythemia:
    electron miscroscopic study, Nouv. Rev. Fr. Hematol.,
    24:19 (1982).
4. R.B. Bolin, T. Okumura, and G.A. Jamieson, Changes in distri-
    bution of platelet membrane glycoproteins in patients with
    myeloproliferative disorders, Am. J. Hematol., 3:63 (1977).
5. B. Boneu, C. Nouvel, P. Sie, C. Caranobe, D. Combes, G. Laurent,
    G. Pris, and R. Bierme, Platelets in myeloproliferative
    disorders, I. A comparitive evaluation with certain
    platelet function tests, Scand. J. Haematol., 25:214 (1980).
6. B.J. Boughton, W.E.N. Corbett, and A.D. Ginsburg, Myleo-
    proliferative disorders: a paradox of in vivo and in vitro
    platelet function, J. Clin. Path., 30:228 (1977).
7. I. Brodsky, Effect of busulfan on oncornavirus-like activity
    in platelets and chromosomes in polycythaemia vera and
    essential thrombocythaemia, J. Natl. Canc. Inst., 59:61
    (1977).
8. C. Caranobe, P. Sie, C. Nouvel, G. Laurent, J. Pris, and B.
    Boneu, Platelets in myeloproliferative disorders, II.
    Serotonin uptake and storage: correlations with mepacrine
    labelled dense bodies and with platelet density, Scand. J.
    Haematol., 25:289 (1980).
9. B. Cooper, A. Shafer, D. Puchalsky, and R. Handin, Platelet
    resistence to prostaglandin D_2 in patients with myelo-
    proliferative disorders, Blood, 52:618 (1978).
10. S. Cortelazzo, M. Colucci, T. Barbui, E. Dini, and N. Semeraro,
    Reduced platelet factor X activating activity. A possible
    contribution to bleeding complications in polycythaemia
    vera and essential thrombocythaemia, Haemostasis, 10:37
    (1981).
11. W. Dameshek, Some speculations on the myeloproliferative
    syndromes, Blood, 6:372 (1951).

12. N. Eche, P. Sie, C. Caranobe, C. Nouvel, J. Pris, and B. Boneu, Platelets in myeloproliferative disorders, III. Glycoprotein profile in relation to platelet function and platelet density, Scand. J. Haematol., 26:123 (1981).

13. E. Epstein and A. Goedel, Hämorrhagische thrombocythämie bei vasculärer Schrumpfmilz, Virch. Arch., 293:233 (1934).

14. M. Fickers and B. Speck, Thrombocythaemia, Acta Haematol., 51:257 (1974).

15. P.G. Frick, Primary thrombocythemia, Helv. Med. Acta, 20:344 (1969).

16. G.F. Gaetani, A.M. Ferraris, S. Galiano, P. Giuntini, L. Canepa, and M. D'Urso, Primary thrombocythemia: clonal origin of platelets, erythrocytes and granulocytes in AGdB/Gd Mediterranean subject, Blood, 59:76 (1982).

17. D. Goldfinger, Long-term plateletpheresis in the management of primary thrombocytosis, Transfusion, 19:336 (1979).

18. L. Gugliotta, A. Guarini, A. Zaccaria, and S. Tura, Platelet function in primary thrombocythemia and chronic myeloid leukemia, Proceedings XVII Congress of the International Society of Hematology, Paris (1978).

19. L. Gugliotta, A. Guarini, C. Timoncini, A. Modiano, M. Aieta, and S. Tura, Valutazione elettronica dei parametri volumetrici delle piastrine nella trombocitemia primitiva e nella porpora thrombocitopenica idiopatica, XXVIII Congresso Nazionale Società Italiana di Ematologia, Bari (1981).

20. L. Gugliotta, G. Cavazzini, M. Baccarani, S. Tura, and The Essential Thrombocythaemia Italian Study Group, The survival of patients with essential thrombocythaemia, (in preparation).

21. L. Gugliotta, G. Biagini, A. Guarini, and P. Preda, Platelet aggregation in essential thrombocythaemia: evaluation by electron microscopy (in preparation).

22. L. Gugliotta, C. Pickering, M. Greaves, and F.E. Preston, Abnormality of platelet membrane glycoproteins in essential thrombocythemia, Thromb. Haemost., 50:216 (1983).

23. F.W. Gunz, Hemorrhagic thrombocythemia: a critical reveiw, Blood, 15:706 (1960).

24. R.M. Hardisty and H.H. Wolff, Haemorrhagic thrombocythaemia: a clinical and laboratory study, Br. J. Haematol., 1:390 (1955).

25. L.A. Harker and C.A. Finch, Thrombokinetics in man, J. Clin. Invest., 48:963 (1969).

26. L.A. Harker, Thrombokinetics, in: "The Platelet," K.M. Brink-
    hous, ed., Williams and Wilkins, Baltimore (1971).

27. S. Holme, M. Simmonds, R. Ballek, and S. Murphy, Comparative
    measurements of platelet size by Coulter Counter, micro-
    scopy of blood smears and light transmission studies.
    Relationship between platelet size and shape, J. Clin. Med.,
    97:610 (1981).

28. P. Kaywin, M. McDonough, P. Insel, and S.J. Shattil, Platelet
    function in essential thrombocythaemia. Decreased
    epinephrine responsiveness associated with a deficiency
    of platelet α-adrenergic receptors, N. Engl. J. Med.,
    299:505 (1978).

29. H. Lander and M.G. Davey, The behaviour and survival of
    platelets in polycythaemia and thrombocythaemia, Aust.
    Ann. Med., 13:207 (1964).

30. J. Lazlo, Myeloproliferative disorders (MPD): myelofibrosis,
    myelosclerosis, extramedullary haematopoiesis, undif-
    ferentiated MPD, and haemorrhagic thrombocythaemia,
    Semin. Hematol., 12:409 (1975).

31. S.M. Lewis, L. Szur, and A.V. Hoffbrand, Thrombocythemia, in:
    "Clinics in Haematology," Vol. 1 (1972).

32. M.L. Mester, L. Szabados, P.R. Guinebault, J. Caen, and S.
    Levy-Toledano, Variations de l'acide sialique plaquettaire
    dans les thrombopathies constitutionelles et acquieses,
    C.R. Acad. Sci. Paris, 273:2157 (1971).

33. S. Murphy, J.L. Davis, P.N. Walsh, and F.H. Gardner, Template
    bleeding time and clinical hemorrhage in myeloproliferative
    disease, Arch. Intern. Med., 138:1251 (1978).

34. Y. Najean, Platelet life span, Ann. Rev. Med., 20:47 (1969).

35. J. Nishimura, S. Okamoto, and H. Ibayashi, Abnormalities of
    platelet adenine nucleotides in patients with myelo-
    proliferative disorders, Thromb. Haemost., 41:787 (1979).

36. M. Okuma and H. Uchino, Altered arachidonate metabolism by
    platelets in patients with myeloproliferative disorders,
    Blood, 54:1258 (1979).

37. F.I. Pareti, A. Capitanio, L. Mannucci, C. Ponticelli, and P.M.
    Mannucci, Acquired dysfunction due to the circulation of
    "exhausted" platelets, Am. J. Med., 69:235 (1980).

38. F.I. Pareti, L. Gugliotta, L. Mannucci, A. Guarini, and P.M.
    Mannucci, Biochemical and metabolic aspects of platelet
    dysfunction in chronic myeloproliferative disorders,
    Thromb. Haemost., 47:84 (1982).

39. F.E. Preston, I.G. Emmanuel, D.A. Winfield, and R.C. Malia, Essential thrombocythaemia and peripheral gangrene, Br. J. Med., 3:548 (1974).

40. N.H. Russel, J. Salmon, J.P. Keenan, and A.J. Bellingham, platelet adenine nucleotides and arachidonic acid metabolism in the myeloproliferative disorders, Thromb. Res., 22: 389 (1981).

41. A.I. Schafer, Deficiency of platelet lipoxygenase activity in myeloproliferative disorders, N. Engl. J. Med., 306:381 (1982).

42. R.H. Spaet, I. Lejnieks, E. Gaynor, and M.L. Goldstein, Defective platelets in essential thrombocythemia, Arch. Intern. Med., 124:135 (1969).

43. S. Tura, Le trombocitemie, Haematologica, 64(Suppl.):179 (1979).

44. L.R. Vasserman, L. Vroman, and G. Gelin, Hémorragies et thromboses an cours de la thrombocythémie essentielle (a propos de 3 observations personelles), Sang, 29:560 (1958).

45. A. Weinfeld, I. Branehog, and J. Kutti, Platelets in the myeloproliferative syndrome, Clin. Haematol., 4:373 (1975).

46. M.M. Wintrob, "Clinical Hematology," Lea and Febiger, Philadelphia (1981).

47. A. Zaccaria, M. Baccarani, L. Gugliotta, and S. Tura, 21 q: a candidate chromosome marker of primary thrombocythaemia, Haematologica, 63:337 (1978).

48. S. Zucker and C.H. Mielke, Classification of thrombocytosis based on platelet function tests: correlation with hemorrhagic and thrombotic complications, J. Lab. Clin. Med., 80:385 (1972).

THE FIBRINOLYTIC ACTIVITY OF

MALIGNANT CELLS

Maria Benedetta Donati,
Grazia Conforti, and
Luciana Mussoni

Laboratory for Hemostasis and
Thrombosis Research
Istituto di Ricerche Farmacologiche
"Mario Negri"
Milan, Italy

INTRODUCTION

The fibrinolytic system is supposed to play a key role in the dissolution of intravascular fibrin deposits and in the prevention of thrombosis. Moreover, several pieces of evidence have emerged recently suggesting that the same system would be of importance in many other physiopathological processes. These include inflammatory responses associated with macrophage activation, cancer cell dissemination, basement membrane metabolism, myogenic differentiation, and even follicle rupture at the time of ovulation or trophoblast implantation, all processes accompanied by tissue remodeling and cell migration. It would thus appear that production of plasminogen activator (PA), the key enzyme triggering fibrinolysis, would be the most widely distributed mechanism used by cells for generating extracellular proteolysis. Several types of cells have indeed been shown able to synthesize and release a PA activity: these include normal cells such as endothelial cells, fibroblasts, smooth muscle cells, macrophages, leukocytes, and pathological cells such as malignant cells. We shall briefly review some of the most attractive pieces of information that have emerged recently regarding the fibrinolytic activity of malignant cells.

TRANSFORMED CELLS

Many tumor tissues are the source of both procoagulant and
fibrinolytic activity, the extent of both being largely independent
from each other, but varying from one tumor to another. There is
no agreement so far whether malignant tissues do indeed produce
higher amounts of fibrinolytic activity than their normal or benign
counterparts. PA synthesis can be stimulated in normal cells by
transforming agents such as viruses, chemical substances, retinoids,
carcinogenic promoters, and physical carcinogens. Cells with a
long history in culture may lose the property to produce PA without
losing tumorigenicity upon reintroduction into the host. Whereas,
the induction of PA production appears a consistent feature of
freshly transformed cells, at least of fibroblasts and of some
epithelial cell lines. Such a phenomenon is not obligatorily
linked to any other phenotypic change, although good correlation
between PA production and loss of anchorage dependence has been
observed. It has very recently been shown that, in avian and mam-
malian fibroblasts, the synthesis of PA is induced by UV radiations
and agents that inflict DNA damage. Under these conditions, PA
generation may accompany the induction of DNA repair enzymes. PA
synthesis would thus be a response, which, like neoplasia, is
limited to cells which retain their capacity for DNA replication.

TUMOR CELLS

The fibrinolytic activity released by most tumors and transformed cells is a serin-protease PA which, in the majority of human
tumors tested, has been found to be identical to urokinase. Based
on this observation, a radioimmunoassay has been set up to detect
the PA release by human ovarian carcinomas, in an attempt to develop
new markers of malignancy. A different type of activator has been
isolated from cells of a human melanoma line. Its characteristics
are similar to those of tissue activators isolated from pig heart
and human bovine uterus. In particular, its very high affinity
for fibrin has offered the basis for developing a new pharmacological
tool for selective and local thrombolysis.

In a recent survey of PA material extracted from a large number
of lung, colon, breast, and prostate carcinomas, the molecular,
immunological, and functional characteristics were of the urokinase
type in almost all cases, even in tumors localized in tissues (such

as lung and colon) which, when normal, contain both types of
activator. While an expanding volume of experimental work is being
produced on the significance of PA as an expression of the transformed phenotype, and on the characterization of PA in tumor tissues,
comparatively little information is available on the significance of
PA produced within experimental (or human) tumors in relation to
their invasive growth and metastasis.

It has been repeatedly proposed that tumors possessing a high
PA activity should be able to digest surrounding fibrin, and penetrate into the circulation much more easily then do tumors with
less PA. However, the evidence for this assumption is still scant.
Some interesting studies have recently been performed on the
fibrinolytic activity of sublines from various murine tumors,
endowed with different metastatic capacity when transplanted in
syngeneic hosts. The PA activity of the highly metastatic sublines
has been found either higher, comparable, or lower than the corresponding non-metastatic lines, depending on the experimental
systems used. In a recent study that we performed on metastatic
variants of a murine, benzypyrene-induced fibrosarcoma produced
a similar type and extent of fibrinolytic activity. Indeed, when
the molecular size of the activator, associated with the cells and
released in the supernatant, was estimated, a single species of
urokinase-like material was found with the same molecular and
antigenic characteristics both in non-metastatic and in highly
metastatic cells. This suggests that production of PA would not
represent one of those cell properties consistently selected in
association with the high metastatic capacity in tumor variants.
The failure to find a clearer correlation between PA and metastatic
capacity in experimental tumors could also be due to the fact that
cell fibrinolytic activity may play opposing roles in the metastic
process, favoring on one hand the detachment of cells from the
primary tumor, and preventing or delaying, on the other hand, the
arrest of the same cells, when they are in the circulation. Indirect support of this assumption is derived from some studies
with antifibrinolytic drugs in experimental tumors, where primary
tumor growth and subsequent spontaneous metastases were decreased,
whereas, lung nodule formation was enhanced. The former effect
is presumably due to a decreased shedding of cells from the primary
tumor, and the latter to microthrombus formation around intravascular trapped tumor cells, a suggested prerequisite for development of metastases from trapped tumor cells.

The factors which will influence in vivo the release and the
availability of PA in the tumor cell microenvironment are still
unknown. In particular, it is not yet clear whether factors known
to stimulate the release of PA from vascular cells (anoxia, veno-
stasis, cathecolamine release) would also be effective on cancer
cell PA. In this context, we have recently made observations on
the modulation of cancer cell PA activity by thrombin, an enzyme
which can be easily generated, at least in traces, by cancer cell
procoagulant activity. Cells from two different murine tumors
(Lewis Lung Carcinoma and mFS6 fibrosarcoma) were found to generate
and release a PA activity which was inhibited by thrombin at such
low concentrations (0.001 U/ml) that the phenomenon could be
considered of physiological relevance. A similar effect of thrombin
had previously been reported on cultured bovine endothelial cells,
and does not result from a direct interaction between thrombin and
the activator (of urokinase-type), but requires the presence of
intact cells in the system.

CONCLUSION

In recent years, a significant amount of knowledge has been
accumulated on the biochemical characteristics of proteolytic
enzymes released from cancer cells and tumor extracts. The development of more sophisticated techniques to culture cells from primary
tumors and metastatic nodules has also made possible to localize
the various enzymatic activities at the cellular level. However,
a gap still exists between observations made in vitro, and events
occurring in vivo in tumor-bearing animals or cancer patients.
Some indirect support for the role of fibrinolysis in malignancy
can be probably obtained from the ongoing pharmacological approaches
with inhibitors and/or activators of fibrinolysis in well defined
experimental systems.

SELECTED BIBLIOGRAPHY

Rather than providing a detailed and scattered list of references on a still developing field of research, we have preferred
to give a selection of few papers, reviews, and books where the
reader can find a larger documentation on specific points of the
subject.

1.   M.B. Donati and A. Poggi, Malignancy and haemostasis, (Annota-
         tion), Br. J. Haematol., 44:173-182 (1980).

2. M.B. Donati, J.F. Davidson, and S. Garattini, eds., "Ma-
    lignancy and the Hemostatic System," Raven Press, New
    York (1981).

3. D.J. Loskutoff, Effect of thrombin on the fibrinolytic activity
    of cultured bovine endothelial cells, J. Clin. Invest.,
    64:329-332 (1979).

4. D.J. Loskutoff and L. Mussoni, Interactions between fibrin
    and the plasminogen activators produced by cultured
    endothelial cells, Blood, 61:in press (1983).

5. G. Markus, et al., Content and characterization of plasminogen
    activators in human lung tissue, Cancer Res., 40:1-6 (1980).

6. L. Mussoni, et al., Plasminogen activator activity of cells
    from primary and metastases of the Lewis Lung Carcinoma,
    in: "Progress in Fibrinolysis," J.D. Davidson, I.M.
    Nilsson, and B. Astedt, eds., Churchill Livingstone,
    Edingburgh, 5:85-89 (1981).

7. H.I. Peterson, Fibrinolysis and antifibrinolytic drugs in
    the growth and spread of tumors, Can. Treat. Rev., 1:213-
    217 (1977).

8. E. Reich, Activation of plasminogen: a widespread mechanism
    for generating extracellular proteolysis, in: "Biological
    Markers of Neoplasia: Basic and Applied Aspects," R.W.
    Ruddon, ed., Elsevier, New York, pp. 491-500 (1978).

ORAL ANTICOAGULANTS IN THE PREVENTION AND

TREATMENT OF VASCULAR DISORDERS

G. Keith Morris

Department of Medicine
University Hospital
Queen's Medical Centre
Nottingham, NG7 2UH

In 1940, Karl Link and his colleagues succeeded in synthesizing dicoumarin. They had identified and were able to manufacture the substance in spoiled sweet clover which had been incriminated as the cause of a severe bleeding disease in cattle. Within one year of its synthesis, dicoumarin was given to man (1), and numerous reports of its clinical application in thrombotic disorders soon followed. But why were clinicians so quick to perceive a therapeutic role for this recently identified cattle poison? The nineteenth century pathologists had emphasized that there were important structural differences between clots and thrombi, and had deduced that the processes of clotting and thrombosis were different, but this message was set aside in favor of the view that thrombosis was clotting in the wrong place (2). It seemed that a substance which could prevent clotting should prevent thrombosis. Moreover, there was a precedent. Heparin, the parenterally administered anticoagulant with a short duration of action, had been discovered earlier, and there was evidence from both animal experiments and clinical application that this anticoagulant could prevent both venous and arterial thrombosis (3). The newly synthesized orally active anticoagulant held out great promise and, initially, there was great enthusiasm for its use. However, the next twenty years were to result in much confusion about the role of oral anticoagulants in the treatment of thrombotic disorders. During this period, important advances in the field of coagulation researach were being made, but the clinical use of dicoumarol and the

vitamin K antagonist drugs which followed tended to run ahead of
these developments.   Control of treatment was a particular problem.
In 1935, Quick et al. (4) developed a test which was thought to measure prothrombin, and even before dicoumarin was synthesized, it was
known that the severity of the bleeding disease in cattle correlated
with the degree of prothrombin deficiency as judged by this test.
Quick's test was, therefore, used to control treatment when
dicoumarin was used clinically, but many workers used modifications
of the original test.   There was uncertainty about the frequency
with which tests should be performed, and it was not fully ap-
preciated that thromboplastins from different sources would influence
the test results.   As a result, degrees of hypocoagulability used
therapeutically varied from center to center and some workers
abandoned control of treatment altogether (5).   That hemorrhage is
the main complication of treatment was widely recognized, but the
frequency with which hemorrhagic complications were reported differed
as did the methods of controlling treatment.   Moreover, the contra-
indications to oral anticoagulant therapy and the effect of drug
interactions that we recognize today were being learned by trial
and error.   Furthermore, during this period the efficacy of treatment was assessed mainly by clinical judgement; an awareness of
the need for prospective control randomized trials with clearly
defined end points in addition to more satisfactory methods of
controlling treatment did not emerge until the late 1950's.   There
is, therefore, a long period in the history of oral anticoagulant
therapy during which uncertainty about the value and safety of the
agents inevitably arose.

Attempts to escape from this period of uncertainty began with
the controlled trials of the 1950's and 60's, but the transition
has not been completed. Now, more than 40 years since the introduction of dicoumarin, we can only say that a consensus has been
reached about the role of oral anticoagulants in the prevention
and treatment of venous thromboembolic disease, but important
aspects of their role in the numerically more important arterial
thrombotic and embolic disorders remain in doubt.   The literature
on these subjects is, of course, extensive and in a brief review
it is inevitable that I shall have to be selective.

ORAL ANTICOAGULANTS IN THE PREVENTION OF VENOUS THROMBOEMBOLISM

The first controlled trial of an oral anticoagulant in the prevention of venous thromboembolism was reported by Sevitt and Gallagher in 1959 (6). They conducted their study on elderly patients with hip fractures, a group with an extremely high risk of venous thromboembolism. Three hundred patients were divided into treated and controlled groups. Allocation to the groups was based on admission on odd and even days of the month. The treated patients were given phenindione in doses sufficient to prolong the one-stage prothrombin time to between two and three times normal, and tests were performed daily for the first three days and then two to three times per week. The mean duration of treatment was five weeks. Patients in the control group received no prophylaxis. The diagnosis of venous thrombosis and pulmonary embolism was based on clinical signs and symptoms, but the authors performed detailed post-mortem studies which included a dissection of the leg veins in patients who died. The effect of treatment was dramatic. Pulmonary embolism did not occur in any patient being treated with phenindione, but occurred in 18% of the control series and was considered to be the cause of death in 10%. Of the control patients who died, 83% were found to have venous thrombosis, but venous thrombosis was found in only 14% of the treated patients who died. There was an excess of postoperative hemorrhagic complications in the phenindione group, but no patient died as a result of treatment. The authors concluded that prophylaxis with phenindione was practicable and safe and would eliminate the risk of pulmonary embolism in patients under its influence.

This trial had its critics (7), and by today's standards there were undoubtedly defects in the trial design. These probably resulted in the control group containing an excess of poor risk patients, thereby exaggerating the ability of treatment to reduce mortality. However, the post-mortem studies in particular provided compelling evidence that treatment could markedly reduce venous thromboembolism.

Salzman et al. (8) studied warfarin sodium in the prevention of venous thromboembolism in elderly patients with hip fractures, and confirmed that treatment could eliminate pulmonary embolism and reduce the frequency of clinically diagnosed deep vein thrombosis, but in that study a significant reduction in mortality was not

observed. Later, Morris and Mitchell (9) also evaluated prophylaxis
with warfarin sodium in these high risk patients in a trial using
a more satisfactory method of randomization. They also demonstrated
the ability of treatment to eliminate pulmonary embolism and bring
about a marked reduction in the frequency of deep vein thrombosis,
as judged by the fibrinogen uptake test during life, and by post-
mortem studies in patients who died. However, treatment failed
to produce a significant reduction in overall mortality. They
found that although treatment was effective in preventing venous
thromboembolism, other causes exerted a high mortality in these
high risk patients. They concluded that treatment should be
administered selectively to patients who could be expected to
survive if pulmonary embolism could be prevented. Despite the
lack of any alternative method of prophylaxis in this group of
patients, prophylactic oral anticoagulant therapy is now used by
only a small minority (2-3%) of orthopedic surgeons in the United
Kingdom (10).

Evidence that oral anticoagulant therapy can prevent venous
thromboembolism in high risk orthopedic patients in whom many risk
factors are operating simultaneously suggests that it should also
be of value in other high risk groups such as patients with
congestive heart failure. However, medical patients have been
studied less extensively, although it is noteworthy that the major
benefit from anticoagulant therapy when used in the early stages
of myocardial infarction seems to stem from a reduction in the
frequency of venous thromboembolism.

ORAL ANTICOAGULANT THERAPY IN ESTABLISHED DEEP VEIN THROMBOSIS

Oral anticoagulant therapy alone or after a period of treatment
with heparin is commonly used in the management of established deep
vein thrombosis. There is certainly strong circumstantial evidence
to indicate that treatment is of value, but no satisfactory trial
has ever been performed to validate its use in this situation.

ANTICOAGULANT THERAPY IN PULMONARY EMBOLISM

Anticoagulation has become the main stay of treatment in
patients with pulmonary embolism but only one prospective, controlled, randomized trial can be sited to support their use. This
trial was performed by Barritt and Jordan in 1960 (11). Patients

with clinical, electrocardiographic, and radiological evidence of
pulmonary embolism were randomized to either a treatment group or
to a non-treatment control group. Treated patients received heparin
(10,000 units 6 hourly) for 36 hours followed by oral nicoumalone
in doses to maintain the prothrombin time at two to three times
normal for 14 days. Only 35 patients were admitted to the
randomized part of this trial (19 control and 16 treated), and
then the study was abandoned because five deaths (attributed to
pulmonary embolism) occurred in the control group, whereas, only
one death (not attributed to pulmonary embolism) occurred in the
treated group. We must remember that this was a pioneering study
and give full credit to these workers, but by today's standards
the design of this trial is open to criticism. In particular,
the numbers included were too small, the criteria used in the
diagnosis of pulmonary embolism were inadequate, and no account
was taken of the prognosis of the underlying illness on which
pulmonary embolism had supervened.

ANTICOAGULANTS IN CORONARY ARTERY DISEASE

The value of anticoagulant therapy in the acute phase of
myocardial infarction has been the subject of extensive study,
but deficiencies and variations in trial design make it very
difficult to draw firm conclusions from any individual study
about its value in this situation. However, in a detailed analysis
of the literature, Chalmers et al. in 1977 (12) concluded that
anticoagulants do confer a small benefit in acute myocardial infarction, and that the benefit is that risk ratio favors their use.
The major part of this small beneficial effect stems from a
reduction in venous thromboembolic complications. In their review,
they pointed out that thromboembolism rates were significantly
lower in the anticoagulated groups in all of the 22 studies in
which the frequency of venous thromboembolism was reported. The
risk of developing deep vein thrombosis in the acute phase of
myocardial infarction is not uniform (13), and patients with
complications, particularly heart failure and dysrhythmias, are
at greatest risk.

Long-term therapy after myocardial infarction to prevent
reinfarction and death has been the subject of dispute for almost
three decades. Many studies (14-25) were performed in the 50's
and 60's, and taken collectively, they suggest that treatment

confers a small benefit. Unfortunately, none of the trials indi-
vidually puts the matter beyond doubt. Interest waned after 1970
when research workers turned to platelet modifying agents and
beta-blockers as possible secondary preventive measures. However,
the report of the Sixty Plus Reinfarction Group in 1980 has reopened
the debate. In this double-blind study, patients aged 60 years
and over, who had been receiving anticoagulants for a period of at
least six months after a myocardial infarction, were allocated at
random to either continue or to cease treatment, i.e., this was
a trial of stopping treatment. Patients who continued on treatment
received either acenocoumarin or phenprocoumonin in doses to maintain
a thrombotest of 5-10%. The control patients were given placebo
tablets and continued to have blood tests. They were assigned
thrombotest values so that their physicians, who did not know the
group to which they belonged, were required to "control" their
treatment. Each group contained 439 patients and the follow-up
period was two years. Two methods of analysis were used. First,
the fate of all patients in each group was analyzed regardless of
any deviation from the trial protocol, and a second analysis was
made excluding patients who had deviated from the protocol for
more than 28 days. Using the first approach, the overall mortality
between the two groups was not significant (51 deaths in the treated
group and 69 deaths in the placebo group - p = 0.07). However,
there was a highly significant reduction in deaths attributed to
reinfarction in the treated group (placebo 27; treated 11 -
p = 0.0005). Using the second method of analysis, a significant
reduction in overall mortality also emerged. This trial is likely
to rekindle interest in anticoagulant therapy as a secondary
preventive measure after myocardial infaraction. New prospective
studies are now needed to settle the questions which the trials of
the 50's and 60's left unanswered, and to determine whether or not
the benefits for elderly patients as suggested by the Dutch study
apply to younger patients.

ANTICOAGULANTS IN CEREBROVASCULAR DISEASE

Anticoagulant therapy in acute stroke probably does more harm
than good. Marshall and Shaw (27) conducted a randomized trial
in patients judged to have sustained a non-embolic, non-hemorrhagic
stroke in which 26 treated patients received phenindione for three
weeks in doses to maintain the prothrombin time at two to three
times normal. The control series of 25 patients received no

treatment. An assessment was made at six weeks when there were
six deaths in the treated group (of which three were attributable
to massive cerebral hemorrhage), and five deaths in the control
group (of which one was attributable to cerebellar hemorrhage).
In this study, the authors attempted to exclude cases of cerebral
hemorrhage and performed cerebral angiography before entering the
patients into the trial. However, cases in which cerebral hemorrhage caused the original stroke could have been mistakenly included,
and the authors emphasized the difficulty in establishing the
underlying pathology (thrombosis, embolism, or infarction) in
the common clinical syndrome of acute stroke.

In trials of long-term anticoagulant therapy, after a clinically non-hemorrhagic cerebrovascular accident no clear benefit has
emerged, and the overall impression is that treatmnet increases
the risk of cerebral hemorrhage and death, particularly in hypertensive patients.

There is a possible role for anticoagulant therapy in the
management of patients with transient cerebral ischemic attacks.
Definitions of a transient ischemic attack vary, but in general
terms, they are short-lived episodes of neurological disturbance
which resolve without any significant residual disability. The
weight of evidence indicates that anticoagulant therapy markedly
reduces the frequency of such episodes and reduces the risk of a
subsequent stroke. However, overall mortality does not appear to
be influenced (32-36). Nonetheless, this is not a unanimous view.
In a small but well designed study, Pearce et al. (37) were unable
to demonstrate any benefit from anticoagulant therapy in patients
with transient ischemic attacks. The question is likely to remain
unresolved because as Genton et al. (38) have calculated: if the
frequency of stroke in patients with TIA is 7% per annum, a trial
involving more than 1,000 patient years would be required to confirm
the value of a drug which could reduce the frequency by 50%.

ATRIAL FIBRILLATION, VALVULAR HEART DISEASE, AND PROSTHETIC HEART
VALVES

The risk of systemic embolization in patients with atrial
fibrillation and rheumatic mitral valve stenosis is well known and
it is generally accepted that anticoagulant therapy will reduce
this risk. Clinical observations and retrospective analyses have

provided strong circumstantial evidence that anticoagulants are
effective in reducing systemic emboli in these patients (39,40),
but no adequately controlled trial has been conducted to confirm
this. The incidence of systemic embolism in untreated patients
has been estimated to be between 3.7% and 8% per patient year
(41,42), but the frequency was only 0.8% in the series of patients
reported by Fleming and Bailey (40). Patients with mitral stenosis
and sinus rhythm, trivial mitral stenosis, or dominant mitral
regurgitation are also at risk from systemic embolization, but
clinicians vary in their attitudes towards prophylactic anticoagulation for these conditions. Similarly, atrial fibrillation secondary
to thyrotoxicosis carries a significant risk of systemic embolization
(43), but whether to treat such patients prophylactically with
anticoagulants remains a matter of clinical judgement and practices
appear to vary widely. In patients with prosthetic heart valves,
anticoagulant therapy does not confer immunity from emboli arising
on or around the valves, but they reduce the risk considerably
(44,45), and the degree of protection may be enhanced by the
simultaneous use of dipyridamole (46).

REFERENCES

1.   H.R. Butt, E.V. Allen, and J.L. Bollman, Preparation from
          spoiled sweet clover (3,3' − methylene-bis-(4hydroxy-
          coumarin) which prolongs coagulation and prothrombin
          time of blood; preliminary reports of experimantal and
          clinical studies, Proc. Staff Meet. Mayo Clin., 16:388
          (1941).
2.   J.R.A. Mitchell, The prevention of thrombosis, in: "Advanced
          Medicine," Pitman Medical, Tunbridge Wells (1978).
3.   G.D.W. Murray and C.H. Best, The use of heparin in thrombosis,
          Ann. Surg., 108:163 (1963).
4.   A.J. Quick, M. Stanley-Brown, and F.W. Bancroft, A study of
          the coagulation defect in hemophilia and in jaundice,
          Am. J. Med. Sci., 190:501 (1935).
5.   R.W. Kistner and G.V. Smith, A ten year analysis of thrombo-
          embolism and dicumarol prophylaxis, Surg. Gynecol. Obstet.,
          98:437 (1954).
6.   S. Sevitt and N.G. Gallagher, Prevention of venous thrombosis
          and pulmonary embolism in injured patients, Lancet, ii:981
          (1959).
7.   G. Eskeland, Prevention of venous thrombosis and pulmonary
          embolism in injured patients, Lancet, i:1035 (1962).

8.   E.W. Salzman, W.H. Harris, and R.W. DeSanctis, Anticoagulation
       for prevention of thromboembolism following fractures on
       the hip, N. Engl. J. Med., 275:122 (1966).

9.   G.K. Morris and J.R.A. Mitchell, Warfarin sodium in prevention
       of deep venous thrombosis and pulmonary embolism in patients
       with fractured neck or femur, Lancet, ii:869 (1976).

10. G.K. Morris, Prevention of venous thromboembolism, Lancet,
       ii:572 (1980).

11. D.W. Barritt and S.C. Jordan, Anticoagulant drugs in the
       treatment of pulmonary embolism. A controlled trial,
       Lancet, i:1309 (1960).

12. T.C. Chalmers, R.J. Matta, H. Smith, and A.M. Kunzler, Evidence
       favoring the use of anticoagulants in the hospital phase
       of acute myocardial infarction, N. Engl. J. Med., 297:1091
       (1977).

13. M.J. Hayes, G.K. Morris, and J.R. Hampton, Comparison of
       mobilization after two and nine days in uncomplicated
       myocardial infarction, Br. Med. J., 3:10 (1974).

14. C.J. Bjerkelund, The effect of long-term treatment with
       dicoumarol in myocardial infarction, Acta Med. Scand.,
       (Suppl.):330 (1957).

15. Medical Research Council, An assessment of long-term anti-
       coagulant administration after cardiac infarction,
       Br. Med. J., 2:837 (1964).

16. R.L. MacMillan, K.W.G. Brown, and D.L. Watt, Long-term anti-
       coagulant therapy after myocardial infarction, Can. Med.
       Assoc. J., 83:567 (1960).

17. J. Clausen, P.E. Andersen, P. Andersen, S. Gruelund, E. Harslof,
       U.H. Andersen, J. Jorgensen, and C. Mose, Long-term anti-
       coagulant treatment after acute coronary occlusion.
       Material with complete comparison with control material,
       Ugeskr. Laeger, 123:987 (1961).

18. B. Harvald, T. Hilden, and E. Lunde, Long-term anticoagulant
       therapy after myocardial infarction, Lancet, ii:626
       (1962).

19. G. Aspenström and K. Korsan-Bengsten, A double-blind study
       of dicoumarol prophylaxis in coronary heart diesease,
       Acta Med. Scand., 176:563 (1964).

20. L.L. Conrad, J.D. Kyriacopaulos, C.W. Wiggins, and G.L. Honick,
       Prevention of recurrences of myocardial infarction. A
       double-blind study of the effectiveness of long-term oral
       anticoagulant therapy, Arch. Intern. Med., 114:348 (1964).

21. A.J. Seaman, H.E. Griswold, R.B. Reaume, and L.W. Ritzman,
     Prophylactic anticoagulant therapy for coronary artery
     disease, J.A.M.A., 189:183 (1964).
22. Veterans Administration Cooperative Study, Long-term anti-
     coagulant therapy after myocardial infarction. A study
     of 747 patients in 15 hospitals, J.A.M.A., 193:929 (1965).
23. E.A. Loeliger, A. Hensen, F. Kroes, L.M. van Dijk, N. Fekkes,
     H. de Jonge, and H.C. Hemker, A double-blind trial of long-
     term anticoagulant treatment after myocardial infarction,
     Acta Med. Scand., 182:549 (1967).
24. R.R.H. Lovell, M.A. Denborough, R.J. Nestel, and A.J. Goble,
     A controlled trial of long-term treatment with anticoagu-
     lants after myocardial infarction in 412 male patients,
     Med. J. Aust., 2:97 (1967).
25. O.J.A. Meuwissen, A.C. Vervoorn, O. Cohen, and F.L.J. Jordan,
     Interim reports on the Utrecht study of long-term anticoagu-
     lant therapy in occlusive coronary artery disease, in:
     "Pathogenesis and Treatment of Thromboembolic Diseases,"
     Proceedings of International Symposium, Basel, Switzerland
     F. Koller, F. Duckert, and F. Streuli, eds., F.K. Schat-
     tauer-Verlag, Stuttgart (1966).
26. Sixty Plus Reinfarction Study Research Group, A double-blind
     trial to assess long-term oral anticoagulant therapy in
     elderly patients after myocardial infarction, Lancet, ii:
     989 (1980).
27. J. Marshall and D.A. Shaw, Anticoagulant therapy in acute
     cerebrovascular accidents. A controlled trial, Lancet,
     1:995 (1960).
28. R.N. Baker, An evaluation of results of anticoagulant therapy,
     in: "Cerebral Vascular Diseases," R.G. Siekert and J.P.
     Whisnant, eds., Grune and Stratton, London (1960).
29. Veterans Administration Report, An evaluation of anticoagulant
     therapy in the treatment of cerebrovascular disease,
     Neurology, 11(Part 2):132 (1961).
30. R.N. Baker, J.A. Broward, H.C. Fang, C.M. Fischer, S.N. Groch,
     A. Heyman, H.R. Karp, E. McDevitt, P. Scheinberg, W.
     Schwartz, and J.F. Toole, Anticoagulant therapy in cerebral
     infarction. Report on cooperative study, Neurology, 12:823
     (1962).
31. A.B. Hill, J. Marshall, and D.A. Shaw, Cerebrovascular disease:
     trial of long-term anticoagulant therapy, Br. Med. J.,
     2:1003 (1962).
32. C.H. Millikan and J.P. Whisnant, Anticoagulant therapy in

cerebral vascular disease. Current Status, J.A.M.A., 166:587 (1958).

33. C.M. Fisher, Anticoagulant therapy in cerebral thrombosis and cerebral embolism. A national cooperative study, interim report, Neurology, 11(Part 2):119 (1961).

34. R.G. Siekert, C.H. Millikan, and J.P. Whisnant, Anticoagulant therapy in intermittent cerebrovascular insufficiency. Follow-up data, J.A.M.A., 176:19 (1961).

35. R.G. Siekert, J.P. Whisnant, and C.H. Millikan, Surgical and anticoagulant therapy of occlusive cerebrovascular disease, Ann. Intern. Med., 58:637 (1963).

36. J. Marshall and E.H. Reynolds, Withdrawal of anticoagulants from patients with transient ischaemic cerebrovascular attacks, Lancet, i:5 (1965).

37. J.M.S. Pearce, S.S. Gubbey, and J.N. Walton, Long-term anticoagulant therapy in transient ischaemic attacks, Lancet, 1:6 (1965).

38. E. Genton, M. Gent, J. Hirsh, and L.A. Harker, Platelet-inhibiting drugs in the prevention of clinical thrombotic disease, N. Engl. J. Med., 293:1174 (1975).

39. S.W. Cosgriff, Chronic anticoagulant therapy in recurrent embolism of cardiac origin, Ann. Intern. Med., 38:278 (1953).

40. H.A. Fleming and S.M. Bailey, Mitral valve disease, systemic embolism and anticoagulants, Postgrad. Med. J., 47:599-604 (1971).

41. N. Coulshed, E.J. Epstein, C.S. McKendrick, R.W. Galloway, and E. Walker, Systemic embolism in mitral valve disease, Br. Heart J., 32:26 (1970).

42. P. Szekely, Systemic embolism and anticoagulant prophylaxis in rheumatic heart disease, Br. Med. J., 1:1209 (1964).

43. J.S. Staffurth, M.C. Gibberd, and S. Ng. Tang Fui, Arterial embolism in thyrotoxicosis with atrial fibrillation, Br. Med. J., 2:688 (1977).

44. M. Akbarian, W.G. Austen, P.M. Yurchak, and J.G. Scannel, Thromboembolic complications of prosthetic cardiac valves, Circulation, 37:826 (1968).

45. G.E. Duvoison, R.O. Brandenburg, and D.C. McGoon, Factors affecting thromboembolism associated with prosthetic heart valves, Circulation, (Suppl. 1):70 (1967).

46. J.M. Sullivan, D.E. Harken, and R. Gorlin, Pharmacologic control of thromboembolic complications of cardiac-valve replacement, N. Engl. J. Med., 279:576 (1968).

PREVENTION OF POSTOPERATIVE VENOUS THROMBOEMBOLISM

BY A NEW LOW MOLECULAR WEIGHT HEPARIN FRACTION

Vijay Kakkar

Thrombosis Research Unit
King's College Hospital Medical School
Denmark Hill, London SE5 8RX

INTRODUCTION

Commercially available heparin represents a heterogenous mass which is the result of a degree of randomness in the biosynthetic process. Therefore, heparin is considered to be a family of straight-chain anionic polysaccharides, more specifically glyco-saminoglycuronan (GAG) sulphate esters of highly variable molecular weight, averaging 9,000 to 15,000 daltons, but extending from perhaps 3,000 to 40,000 daltons (1,2). The anticoagulant action of heparin results from its ability to bind and activate antithrombin III, a heparin cofactor. Antithrombin III inhibits several activated coagulation factors in addition to thrombin, including factors IXa, Xa, XIa, and XIIa; all those inhibition reactions are greatly accelerated in the presence of heparin (3-8). Studies on heparin fractions of different molecular weights obtained by gel filtration, have shown a molecular size dependency of the anticoagulant activity (9-13). The potentiation of the inhibition of factor Xa in the plasma system increases with decreasing molecular weight (9-11). In contrast, measurements by multiple-role clotting assay such as activated partial thromboplasint time (APTT) decreases with decreasing molecular weight (11-14).

A preliminary study has reported that a single daily injection of a low molecular weight heparin fraction is a convenient way of preventing deep vein thrombosis (15). The present study was

designed to find out whether this regimen of heparin prophylaxis
was also effective in preventing postoperative pulmonary emboli
in high risk patients undergoing major abdominal surgery.

PATIENTS AND METHODS

     Five hundred and fifty patients over the age of forty, under-
going a variety of elective major surgical procedures, were included
in this study. All received a written and oral description of the
study protocol, and gave their consent to participate. A major
operation was considered to be one which was performed under
general anesthesia, lasted for approximately an hour, and required
a postoperative hospital stay of at least seven days. Patients
were excluded from this study if they had clinical signs of recent
deep vein thrombosis, were to undergo operations on the thyroid
gland, or if they had already been on prophylactic anticoagulant
therapy at the time of admission to the hospital. The details of
age, sex, type of operation performed, and other risk factors
likely to influence the incidence of deep vein thrombosis were
recorded on a specially designed proforma suitable for computer
analysis.

Low Molecular Weight (LMW) Heparin Fraction

     Commercial preparation of pig mucosal heparin was used for
preparing a new LMW heparin fraction (CY 216 Laboratoire Choay –
Paris). This fraction was (16) extracted with a mixture of ethanol
and water, using the technique previously described and purified.
The preparation obtained was characterized using physical (^{13}C-
nuclear magnetic resonance spectroscopy) and chemical techniques.
Compared with the starting material, this LMW heparin fraction
contained a higher percentage of non-sulphated uronic acid,
particularly glucuronic acid (about 30% of total uronic acids) as
well as a higher percentage of N-acetyl-glucosamine, whereas,
the content in sulphate groups was lower (17). The molecular
weight of CY 216 was estimated by High Performance Liquid Chromatography, and found to be approximately 5,000 daltons (17). The
mean molecular weight of commercial heparin is of 16,000 daltons.
The anticoagulant activities of the two preparations were assessed
using APTT (18), and Yin & Al antifactor Xa clotting assay (19).
This LMW heparin fraction had a specific activity of 200 U/mg in
antifactor Xa assay, and 50 U/mg in APTT assay. The ratio between

Table I. Sex, Age, and Weight Distribution

Sex	No.	%
M	185	36.9
F	316	63.1
Age (yr)		
40-60	241	48.1
60-80	234	46.2
>80	26	5.7
Weight (Kg)		
< 40	10	1.9
40-60	198	39.5
61-80	219	43.7
>80	74	14.7

both types of activities in tests using human plasma was thus 4.
Calciparine had a specific activity of 170 U/mg in both assays.
CY 216 was supplied in a single dose phial prepared under approved
pharmaceutical conditions, containing 1,850 APTT units (75,000
anti-Xa units) in 0.3 ml. Results of tests for sterility and
pyrogens were satisfactory.

Administration of LMW Heparin Fraction

Each patient received a single daily injection of 1,850 APTT
units of LMW heparin, starting two hours before surgery, and
continuing daily for seven days after operation.

ASSESSMENT

Deep Vein Thrombosis

Deep vein thrombosis was detected by a radioactive fibrinogen
uptake test. Each patient received 100 uci of ^{125}I-fibrinogen the
day before surgery. Their legs were scanned using a "Logic 121"
timer/scaler. Counting was performed according to our own
technique (20). DVT was diagnosed if the counts at any site differed by 20% or more from those at an adjacent point on the same

leg or the same position on the opposite limb, and if this difference persisted or increased in the subsequent twenty-four hours
(20), venography was performed (21) whenever a positive scan
developed in a patient.

Pulmonary Embolism

The frequency of clinically suspected pulmonary embolism was
recorded, and any confirmatory evidence of this diagnosis of chest
x-ray, ventilation/perfusion lung scanning was also noted. If the
patient died, the pathologist was requested to record the disease
or condition directly leading to death. Pulmonary trunk, main
pulmonary artery, lobor and segmental arteries were examined for
deep vein thrombosis, and if thrombi were present, their location
was noted.

Blood Loss

The surgeon assessed blood loss during operation as excessive
or not. Surgical wounds were examined every day for hematoma
formation. The hemoglobin value before and a week after surgery,
and blood transfusion requirements were also noted. Heparin
injection site was examined for local hematoma formation or bruising.
Such a reaction was recorded as being greater or lesser than 0.5
cm in diameter.

RESULTS

Five hundred and fifty patients were admitted to this study;
49 patients were excluded from analysis for the following reasons:
operation was cancelled in 15, procedure performed did not fulfill
the criteria of a major surgical operation in 18, and in 16 the
radioactive fibrinogen uptake test was not performed according to
the protocol, the limbs were not scanned for 7 days as the patient
was discharged from hospital on the 5th or 6th postoperative day.
Thus there were 501 patients in whom the protocol had been cor-
rectly followed. Their age, sex, and weight distribution is shown
in Table I, and the operations performed in Table II.

Total Frequency of DVT

DVT developed in 17 (3.5%) of 501 patients investigated; in

Table II. Operations Performed on 501 Patients Receiving LMW Heparin Fraction

	No. of Patients
Gastric	97
Esophagectomy	8
Total and partial gastrectomy	35
Vagotomy and drainage	32
Gasteroenterostomy	17
Fundoplication and hiatus hernia repair	5
Biliary	140
Cholecystectomy	99
Cholecystectomy and exploration of common bile duct	19
Cholecystoenterostomy or choledoduedenostomy	11
Partial pancreatectomy or drainage	3
Drainage of liver or subphrenic abscess	4
Splenectomy	4
Large Bowel	105
Total and partial colectomy	58
Anterior resection of rectum	17
Abdomino-perineal resection of rectum	8
Ilio-colic anastomosis or closure of colostomy	22
Urological	32
Nephrectomy and uretrectomy	8
Ureteric implant	3
Cystectomy	3
Pyelolithotomy	4
Transurethral resection of prostate	14
Gynecological	85
Hysterectomy ± salphingoohrectomy	75
Colosuspension	7
Ovarian cystectomy	3
Miscellaneous	42
Laparotomy and biopsy	25
Radical or simple mastectomy	12
Small bowel resection or bypass	5
Total	501

Table III. Risk Factors Predisposing to Postoperative DVT

Risk Factors		No. of Pts.	
Previous PE		4	(0)
Previous DVT	– Rt. leg	15	(1)
	– Lt. leg	6	(0)
	– Both	5	(0)
Long bone fracture	– Rt. leg	9	(1)
	– Lt. leg	4	(0)
	– Both	2	(0)
Varicose veins	– Rt. leg	21	(2)
	– Lt. leg	21	(1)
	– Both	69	(2)
Healed venous ulcer	– Rt. leg	6	(0)
	– Lt. leg	2	(0)
	– Both	3	(0)
Contraceptive pill		13	(0)

none were thrombi bilateral. In each of the 17 limbs, thrombosis was detected in the calf, and the process did not extend to involve the more proximal veins. In 7 patients, thrombi were detected between the 3rd and 6th postoperative day, in 6 between the 7th and 10th day, and in 4 after the 10th day. In 9 (1.7%) of 17 patients, presence of thrombi was confirmed by phlebography.

DVT and Risk Factors

One hundred and ninety-two patients were operated for malignant disease, and 12 (6.25%) of these developed DVT; of 309 patients operated for benign disease, thrombi were detected in 5 (1.6%). Eighteen patients developed Gram-negative intraperitoneal infection, and thrombi were detected in 2 (1.11%). Thrombi were detected in 15 (3.1%) of 485 patients who did not develop such complications during the postoperative period. The frequency of DVT in relation to other known risk factors is shown in Table III and Table IV.

Table IV. Days between Admission, Operation, Length of Hospital Stay, and DVT

		No. of Patients	Mean	Standard Deviation	Standard Error	Significance
Days between Admission & Operation	No DVT	484	6.034	9.199	0.425	p = 0.15
	Postop DVT	17	9.375	7.856	1.064	
Length of Hospital Stay	No DVT	484	18.257	13.188	0.624	p = <0.03
	Postop DVT	17	26.583	12.064	3.483	

Table V.   Causes of Death in Patients who Had
           Autopsy Examination

Bronchopneumonia	4
Acute renal failure	2
Septicemia	3
Carcinomatosis	6
Acute int. obstruction	1
Hemorrhage	1
Ruptured aortic aneurysm	1
Total	18

Total Mortality Rate

Of 501 patients included in this study, 22 (4.3%) died during
the postoperative period, one on operative day, 4 on the 1st postoperative day, and the others as follows: 1 on the 6th postoperative
day, 2 on the 8th, 1 on the 9th, 2 on the 11th, 1 on the 14th, 16th,
18th, 19th, 27th, 32nd, 43rd, 62nd, and 87th days. Autopsy was
performed in 18 of these, in 4 patients permission for a post-mortem
examination was not granted by the relatives. The cause of death
as determined at autopsy in 18 patients is shown in Table V.

Frequency of Non-Fatal Pulmonary Embolism

Two patients developed clinical features suggestive of pulmonary
embolism, on the 3rd and 8th postoperative days. A normal chest
x-ray and mismatched ventilation/perfusion lung scan confirmed the
presence of emboli in basal segments of the right lung in both
instances. They received 30,000 units of heparin intravenously for
a period of 7 days, and recovery was uneventful.

Operative and Postoperative Bleeding

Blood loss during surgery was considered to be excessive in
32 (6.3%) of 501 patients; in 5 of these, prophylaxis had to be
discontinued. In another 19 (3.2%) patients, blood loss was also
considered to be excessive during the postoperative period.

Table VI. Wound Hematoma vs. Site of Operation

Wound Hematoma	Site of Operation Abdominal	Pelvic	Extra Abdominal	Total
No	388	131	13	482
Yes	5 (1.4%)	14 (10.6%)	0	19 (3.8%)
Total	393	145	13	501

One hundred and forty-three patients received blood transfusion, the average amount transfused being 1,017 ml. The fall in the postoperative hemoglobin was 0.62 g/ml. In 19 (3.2%) patients, subcutaneous hematoma developed at the site of abdominal incision, in 18 hematoma resolved spontaneously, while one required evacuation. The relationship between the frequency of wound hematoma and the site of operation is shown in Table VI. A local hematoma of greater. than 0.5cm in diameter at the injection site was also observed in 8 (1.5%) of 501 patients, of less than 0.5cm in 10 (2.0%). Minor bruising of the injection site occurred frequently. In none of the patients was prophylaxis discontinued because of local pain.

DISCUSSION

A preliminary report had suggested that LMW heparin fragment may be a convenient way of preventing deep vein thrombosis in high risk patients. However, it failed to provide evidence whether such heparin prophylaxis was also effective in preventing postoperative fatal pulmonary emboli, since the number of patients investigated was rather small. The present study provides and answer to this crucial question. Of 501 patients included in the present study, 22 patients died during the postoperative period, and autopsy was performed in 18; pulmonary emboli could be demonstrated in none. In 4 patients, permission for a post-morted examination was not granted; the probable cause of death in these patients was generalized carinomatosis (2 patients) and chronic renal failure (2 patients). The patients included in the present study truly represent a high risk population, since 234 (46.2%) were between the age of 60-80 years, and 192 (38.3%) were operated for malignant disease. In a multicenter trial where 2,076 similar patients not receiving any prophylaxis were investigated, pulmonary emboli were demonstrated at autopsy in 22 (30.5%) of 72 patients (22). Thus, it appears that the regimen of LMW heparin used in this study is highly effective in preventing death due to fatal pulmonary emboli.

The present study also provides further evidence that LMW heparin regimen is effective in reducing the incidence of isotopic deep vein thrombosis, since in only 17 (3.5%) of 501 patients such thrombi were detected. Several studies have now demonstrated that approximately 25-30% of the high risk patients would be expected to develop isotopic thrombi during the postoperative period when routine prophylaxis is not employed (23-27).

Patients who have surgery for malignant disease are considered to have a particularly high risk of developing venous thrombo-embolism (22,28,29), and a number of patients having operations for malignancy were included in several trials of heparin prophylaxis, in general surgical patients. Analysis of data of such patients indicate that low dose heparin, using commercially available preparations, is effective in protecting such patients; isotopic deep vein thrombosis occurred in 49% of patients with malignancy, who did not receive prophylaxis, compared to 13% in those who did receive low dose heparin for protecting such patients, since in the present study only 12 (6.25%) of 192 patients operated for such a disease process developed postoperative thrombi.

Few physicians would deny that venous thrombosis, although very common, is mostly a benign disease. Using phlebography and the ^{125}I-labelled fibrinogen test, it has been shown that in surgical patients the majority of thrombi form in the calf veins. A surprisingly high proportion of these undergo spontaneous lysis and only in something like 20% of patients do these thrombi extend more proximally from the calf into the popliteal, femoral, and iliac veins (30). In this group with extending thrombosis, pulmonary embolism occurs in almost 50%, but only a very small proportion prove fatal (30). The effect of low dose heparin prophylaxis on the extension of venous thrombosis was evaluated in four of the larger studies of patients undergoing elective abdominal surgery, including more than 3,000 patients (22,31,32). Of 1,631 patients in the control group, thrombi were detected in 380, while extension of thrombosis occurred in 99 (i.e., 60%). In contrast, out of 1,485 patients included in the low dose heparin group, thrombi were detected in 95 and extension occurred in only 9 (i.e., 0.6%). The difference in the frequency of extending thrombi between the two groups was of high statistical significance, not only in the aggregate of those four studies, but in each of the individual trials as well (22,31,32). LMW heparin fragment prophylaxis seems to be even superior than low dose heparin, since in none of the 17 patients who developed isotopic deep vein thrombosis did thrombi extend to involve the more proximal veins.

The main limitation for using anticoagulants in the prevention of thromboembolic disease in surgical patients is the risk of hemorrhage. The only definite criterion which one can use to evaluate this risk is the frequency of wound hematoma formation. In the two reported studies, in which large numbers of patients were investigated, a significant difference was observed in the development of wound hematoma between the number of patients receiving heparin and their control counterparts (15,32). In the multicenter trial (15), 158 (7.7%) of 2,076 patients who received low dose heparin prophylaxis developed wound hematoma. The regimen of LMW heparin fragment prophylaxis used in the present study appears to be safer since only 19 (3.2%) of 501 patients developed this complication. Bruising at the injection site has also been observed, when either the calcium or sodium salt of commercially available heparin has been used. In a comparative trial of 266 subcutaneous injections (133 each of sodium and calcium heparin), sodium heparin caused local bruising greater than 0.5cm in 47% and none at all in 46%, while calcium heparin produced local hematoma or bruising of over 0.5cm in diameter in 3.0%, under 0.5cm in 22%, and no bruising in 75% (32). LMW heparin fragment seems to be better tolerated, since in the present study, a local hematoma or bruising of greater than 0.5cm in diameter at the injection site was observed in only 8 (1.5%) of 501 patients, bruising of less than 0.5cm in only 10 (2.0%), while in 96.5% of patients, no such local reaction was observed. In one of the patients was prophylaxis discontinued because of local pain.

The value of low dose commercially available heparin in prophylaxis of postoperative venous thromboembolism can be no longer seriously disputed. However, before any therapy can be widely adopted, it is essential that it can be easily administered, readily available, of low cost, and above all, of minimal risk when the patient is subjected to major tissue trauma. The results of the

present study indicate that a single daily injection of 1,850 APTT
units of CY 216 LMW heparin fragment is a convenient way of preventing postoperative venous thromboembolic complications in high
risk patients undergoing major tissue trauma. The results of the
present study indicate that a single daily injection of 1,850 APTT
units of LMW heparin is a convenient way of preventing postoperative
venous thromboembolic complications in high risk patients undergoing
major abdominal surgery. This regimen not only reduces the frequency
of hemorrhagic side effects, but also overcomes the need for
frequent drug administration, and thus may lead to a better acceptance by the patients and surgeons.

REFERENCES

1.   J.A. Cifonelli, The relationship of molecular weight, and
          sulfate content and distribution to anticoagulant activity
          of heparin preparation, Carbohydr. Res., 37:145-154 (1974).
2.   E.A. Johnson and B. Malloy, The molecular weight range of
          mucosal heparin preparation, Carbohydr. Res., 51:119-123
          (1976).
3.   L.H. Lam, J.E. Silbert, and R.D. Rosenberg, The separation of
          active and inactive form of heparin, Biochem. Biophys. Res.
          Commun., 69:570-576 (1976).
4.   R.D. Rosenberg, Actions and interactions of antithrombin and
          heparin, N. Engl. J. Med., 292:146-148 (1975).
5.   U. Abildgaard, Inhibition of the thrombin-fibrinogen reaction
          by heparin and purified cofactor, Scand. J. Haematol.,
          5:440-445 (1968).
6.   J.S. Rosenberg, P.W. McKenna, and R.D. Rosenberg, Inhibition
          of human factor IXa by human anti-thrombin, J. Biol. Chem.,
          250:8883-8888 (1975).
7.   B. Osterud, M. Miller-Andersson, U. Abildgaard, and H. Prydz,
          The effect of anti-thrombin III on the activity of the
          coagulation factor VII, IX, and X, Thromb. Haemostas., 35:
          295-301 (1976).
8.   D.A. Lane, I.R. MacGregor, R. Michalski, and V.V. Kakkar,
          Anti-coagulant activities of four unfractionated and
          fractionated heparins, Thromb. Res., 12:257-271 (1978).
9.   L.O. Andersson, T.W. Barrowcliffe, E. Holmer, E.A. Johnson, and
          G.E.C. Sims, Anticoagulant properties of heparin fractionated
          by affinity chromatography on matrix bound anti-thrombin III
          and by gel filtration, Thromb. Res., 9:575-583 (1976).

10. P.S. Damus, S. Hick, and R.W. Rosenberg, Anticoagulant action
       of heparin, Nature, 246:355 (1973).
11. T.C. Laurent, A. Tengblad, L. Thurnberg, M. Hook, and U.
       Lindahl, The molecular weight dependence of the anti-
       coagulant activity of heparin, Biochem. J., 175:691-701
       (1978).
12. L.O. Andersson, T.W. Barrowcliffe, E. Holmer, E.A. Johnson,
       and G. Soderstrom, Molecular weight dependency of the
       heparin potential inhibition of thrombin and activated
       factor X. Effect of heparin neutralisation in plasma,
       Thromb. Res., 15:531-541 (1979).
13. E.A. Johnson, T.B.L. Kirkwood, Y. Stirling, L.J. Perez-Requejo,
       G.I.C. Ingram, D.R. Bangham, and M. Brozovic, Four heparin
       preparations. Anti-Xa potentiating effect of heparin after
       subcutaneous injection, Thromb. Haemostas., 35:586-591
       (1976).
14. T.W. Barrowcliffe, E.A. Johnson, C.A. Eggleton, G. Kemball-Cook,
       and D.P. Thomas, Anticoagulant activities of high and low
       molecular weight heparin fractions, Br. J. Haematol.,
       41:583 (1979).
15. An International Multicentre Trial, Prevention of Fatal
       Postoperative pulmonary embolism by low doses of heparin,
       Lancet, ii:45-51 (1975).
16. J. Choay, J.C. Lormeau, M. Petitou, P. Sinay, B. Casu, P.
       Poreste, G. Torri, and G. Gatti, Anti-Xa active heparin
       oligosaccharides, Thromb. Res., 18:573-578 (1980).
17. J. Choay, Personal communication.
18. R.M. Hardisty and G.I.C. Ingram, "In Bleeding Disorders,"
       Oxford (1965).
19. E.T. Yin, S. Wessler, and J.V. Butler, Plasma heparin: a unique
       practical, submicrogram-sensitive assay, J. Lab. Clin. Med.,
       81:293-310 (1973).
20. V.V. Kakkar, A.N. Nicolaides, J.T.C. Renney, J.R. Friend, and
       M.D. Clarke, ^{125}I-fibrinogen test adapted for routine
       screening, Lancet, i:540-542 (1970).
21. V.V. Kakkar, The problems of thrombosis in the deep veins of
       the legs, Ann. Roy. Coll. Surg. Engl., 45:257-276 (1969).
22. A.S. Gallus, J. Hirsch, T.J. Tuttle, R. Trebilcock, S.E.
       O'Brien, J.J. Carroll, J.H. Minden, and S.M. Hudecki,
       Small subcutaneous doses of heparin in prevention of venous
       thrombosis, N. Engl. J. Med., 288:545-549 (1963).
23. V.V. Kakkar, J. Spinder, P.T. Flute, T. Corrigan, D.P. Fossard,
       R.Q. Crellin, S. Wessler, and E.T. Yin, Efficacy of low

doses of heparin in prevention of deep vein thrombosis after major surgery; a double blind trial, Lancet, ii: 101–106 (1972).

24. A.N. Nicolaides, S. Desai, J.N. Douglas, G. Fourides, P.A. Dupont, J.D. Lewis, H. Dodsworth, K.J. Luck, and C.W. Jamieson, Small doses of sodium heparin in preventing deep venous thrombosis after surgery, Lancet, ii:890–893 (1972).

25. J. Kill, F. Axelson, and D. Andersen, Prophylaxis against postoperative pulmonary embolism and deep vein thrombosis by low dose heparin, Lancet, i:1115–1116 (1978).

26. V.V. Kakkar, C.T. Howe, A.N. Nicolaides, J.T.G. Renney, and M.B. Clarke, Deep vein Thrombosis of the leg. Is there a high risk group? Am. J. Surg., 120:527–531 (1970).

27. V.V. Kakkar, C.T. Howe, C. Flanc, and M.B. Clarke, Natural history of deep vein thrombosis, Lancet, ii:230–233 (1969).

28. A.N. Nicolaides, D. Irving, M. Pretzell, P. Dupont, J. Lewis, S. Desai, J.N. Douglas, V.V. Kakkar, and E.S. Field, The risk of deep vein thrombosis in surgical patients, Br. J. Surg., 60:312–317 (1973).

29. E.E. Sproul, Carcinoma and venous thrombosis, Am. J. Cancer, 24:566–570 (1938).

30. V.V. Kakkar, The problems of thrombosis in the deep veins of the leg, Ann. Roy. Coll. Surg. Engl., 45:257–276 (1969).

31. S.A. Gallus, S. Hirsh, R.S. Tuttle, R. Trebilcock, and S.E. O'Brien, Small subcutaneous doses of heparin in the prevention of venous thrombosis, N. Engl. J. Med., 288: 545–549 (1973).

32. A.N. Nicolaides, P.A. Dupont, and S.A. Desai, Small doses of subcutaneous sodium heparin in preventing deep vein thrombosis after major surgery, Lancet, ii:890–894 (1972).

DISSEMINATED INTRAVASCULAR COAGULATION

P. Citarrella and A. Cajozzo

Patologia Medica I
Università di Palermo
Palermo, Italy

INTRODUCTION

During the past decade, diffuse intravascular coagulation (DIC) has been one of the most intensively studied subjects in hematology. DIC is a dynamic pathologic process caused by the entrance of foreign material into the circulation, or by exposure to abnormal surfaces. The intravascular coagulation is triggered by activation of the clotting cascade with resultant generation of excess thrombin within the vascular system that leads to further activation of the coagulation system, shortened survival of certain hemostatic elements, deposition of fibrin in the microcirculation, and activation of the fibrinolytic system. Decreased levels of hemostatic factors, secondary fibrinolysis, and exhaustion of the physiologic inhibitors (AT III and SRE) may result in clinically excessive bleeding, particularly when the patient's blood vessels have been damaged. In contrast, fibrin deposition may cause micro- or macro-thrombosis, and occasionally results in a fragmentation-type hemolytic anemia (Fig. 1).

FACTORS WHICH CAUSE DIC

DIC should not be considered as a disorder in itself, but as a process that is caused by a great variety of underlying diseases (Table 1). Many conditions, including common complications of pregnancy and delivery, acute infection, neoplasms, disorders of the hematopoeitic system, vascular disorders, massive tissue injury,

Table 1. Mechanisms by which DIC is Initiated

Intrinsic Pathway	Extrinsic Pathway	Common Pathway	Platelet Adhesion and Aggregation
XII	VII	X	Endotoxins
Endotoxins	Massive trauma	Mucinous material	Massive endothelial injury
Bacterial fragments	Burns	Proteolytic enzymes	Giant hemangiomas
Antigen-antibody complexes	Obstetric complications	II-I	Antigen-antibody complexes
Collagen	Promyelocytic leukemia	Proteolytic enzymes	Rickettsial and viral infections
	Neoplasms		Major hemolytic trans- fusion reactions
	Giant hemangiomas		Primary and secondary thrombocytosis
	Intravascular hemolysis		Particulate matter
	Endotoxins		
	Surgery of prostate, pancreas, and lung		

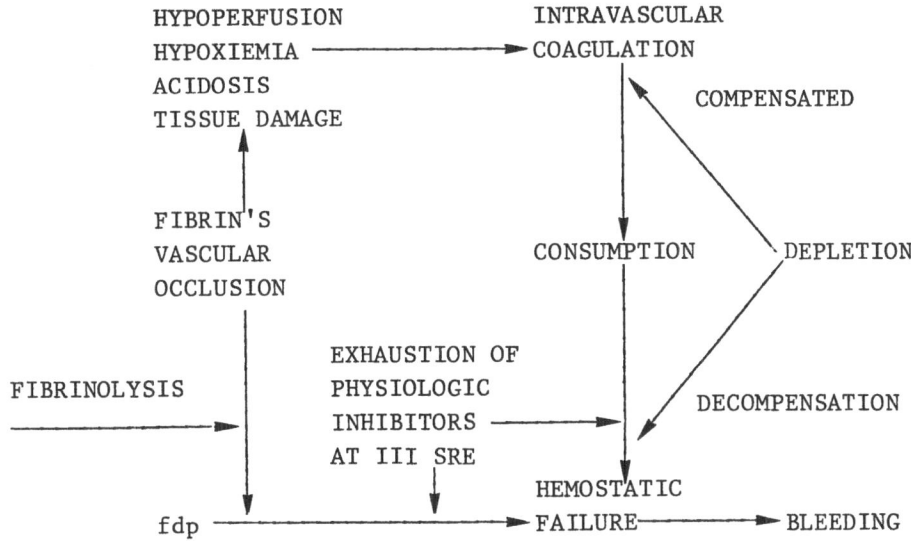

Fig. 1. Physiopathology of DIC.

and others are able to activate the coagulative process causing
intravascular fibrin formation especially in the microcirculation
(7,8,12,14,18,20,22). However, in most forms of DIC the initiating
factors are multiple and interrelated. The mechanisms by which
DIC is initiated are numerous, but they all have in common the
capacity, in terms of either the magnitude or the duration of the
activating stimulus, to exceed or overwhelm normal compensatory
process. The activation of DIC involves the normal mechanism
of hemostasis: the process of platelet adhesion and aggregation,
the contact-activated (intrinsic) and tissue-activated (extrinsic)
pathways of coagulation (Table 1). The entry of procoagulant tissue
extracts into the blood from endogenous sites is presumably a
contributing factor in most forms of DIC, and is of major pathogenetic importance in cases associated with obstetric complications,
neoplasms, and disorders of the hematopoietic system (2,5,7,9,11,17).
Tissue extracts interact with factor VII and activate the extrinsic
pathway of coagulation.

Frequently DIC is accompanied by septicemia caused by bacteria
that possess potent endotoxin. This has led to the intensive study
of the effect of endotoxin on the hemostatic apparatus. Endotoxins
produce activation of complement, intrinsic pathway, direct endothelial injury, platelet and leukocyte aggregation (Fig. 2) (12).
Miscellaneous activating stimuli (antigen-antibody complexes,

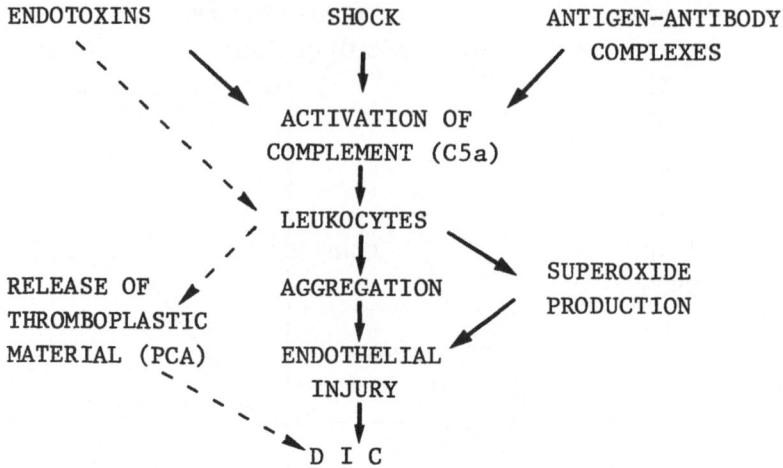

Fig. 2.   Role of leukocytes in pathogenesis of DIC.

snake venoms, etc.) may initiate DIC with activation of factors
XII, X, or prothrombin. Shock, hypoperfusion, and hypoxemia may
favor the development and may perpetuate DIC. In order for DIC
to occur, there must necessarily be a simultaneous exhaustion of
the physiologic inhibitors which control the homeostasis of the
coagulative process. In particular AT III and reticuloendothelial system (Fig. 1) (1,3,10,13,18).

COAGULATION FACTORS AND CLINICAL MANIFESTATIONS

    DIC constitutes a model of accelerated turnover of various
coagulation factors. Numerous studies have demonstrated accelerated
turnover rates for platelets, fibrinogen, and prothrombin. Thrombin
is persistently elaborated and fibrin is formed in circulating blood.
Fibrinogen, various other coagulation factors, and platelets are
utilized. The fibrinolytic system is activated and large amounts
of fibrinogen products are produced. These further impair the
hemostatic function (16,18,20,22). Bleeding, shock, and vascular
occlusion commonly supervene and produce profound alterations in
the function of various organ systems.

    The clinical manifestations of DIC are extremely variable, not
only from patient to patient, but also in the same patient from
time to time. These differences depend on the features of the
underlying disease, the intensity and duration of activation of
blood coagulation and the rates of thrombin formation, the state

Table 2. Diagnosis of DIC

Clinical Features	Hematological Findings	Useful Hematological Tests	Coagulation Studies	Other, Less Useful, Time Consuming Tests
Symptoms of under-lying disease	Thrombocytopenia	Hemoglobin level	Fibrinogen level	FPA (after bolus of heparin and Δ -FPA)
Bleeding	Fragmentation-type anemia	Packed cell volume	fdp	F XIIIRAg:A
Thrombosis	Hemoglobinemia	White cell count	Paracoagulation tests	HMWFC
Intravascular hemo-lysis	Hemoglobinuria	Platelet count	AT III	Kallikrein
Shock		Blood film	PTT, TT, PT	PF4, β-TG
				Leukocyte protein L1
				C3, C4
				Factor VIII
				Platelet & fibrino-gen survival
				Plasminogen, Eu-globulin lysis
				Fibronectin

of the fibrinolytic system, the rate of blood flow and the level
of liver function, the bone marrow, and the macrophage system.
However, even within this wide range of possible symptoms, in some
patients DIC will be of little clinical significance, and the
diagnosis is made on the basis of laboratory data. In other
patients, DIC is so severe that symptomatology of the underlying
disease is obscured by serious hemorrhaging, hypoperfusion, and
thrombosis.

DIAGNOSIS OF DIC

Because of its extremely variable clinical presentation, it is
not surprising that diagnostic and laboratory criteria for DIC are
not yet well defined (Table 2). It seems clear, however, that
frequency of diagnosis of this hemostatic disorder depends essentially on the physician's awareness of the possibility of DIC's
association with certain disease states and laboratory data (8,18,
20,22).

Laboratory data used to reach a diagnosis of DIC are often
arbitrary and many laboratory values are considered without reference
to the underlying disease of the patient or organ dysfunction.
For example, a recent study showed that the fibrinogen level,
usually thought to be of great importance in diagnosing DIC, was
within normal limits in 60% of the time. In contrast in severe liver
diseases complicated by DIC, a low fibrinogen level is a result of
reduced production and increased consumption (8). Similar problems
may arise when other criteria that can be influenced by the underlying conditions are used to reach a diagnosis, for example,
thrombocytopenia during an acute leukemia or in the presence of
splenomegaly or sepsis, low factor V or low AT III with impaired
liver function, factor VIII levels in pregnancy or in cases of
infection, elevated levels of fibrin degradation products (fdp)
in severe deep hemorrhage, serous effusions, exudates, or primary
fibrinolysis. Because of the variability of the afore-mentioned
criteria, the laboratory diagnosis of DIC is not always assured.

In our opinion, it is above all necessary to document that
thrombin has reacted with the fibrinogen molecule, breaking off the
peptide fragments A and B and forming the fibrin monomer which, if
present in critical concentrations, can form a soluble complex with
fibrinogen. This reaction can be demonstrated by paracoagulation

Table 3. Treatment of Acute DIC

---

1 - Treatment of underlying dis-
    ease

2 - Replacement therapy

3 - Heparin

4 - Other therapeutic measures
    for shock, ARDS, renal fail-
    ure, etc.

---

tests with ethanol or protamine sulfate which may, however, be
negative in cases of a very low fibrinogen concentration. Other
more recent, less rapid or repeatable, and more complex methods
demonstrate the products of the direct reaction of thrombin on
fibrinogen (peptide fragment A levels) (11,21). There is no one
single test which can be used to confirm or exclude the diagnosis
of DIC. However, the combination of a low platelet count, positive
paracoagulation tests, altered levels of fibrinogen, fdp, and AT III,
evaluated in the framework of the underlying condition, seem to be
of great help as indexes in making a diagnosis of DIC which is in
agreement with the clinical data (5,8,17,18,20).

TREATMENT OF DIC

     The treatment of acute DIC represents one of the most difficult
problems in the field of medicine because of the difficulty encountered in interpreting the symptoms and hematological data, the
gravity of the clinical condition, and the rapidity of evolution
of the syndrome (Table 3). There are several questions that must
be asked to determine whether replacement of depleted hemostatic
factors is necessary, and interruption of DIC with heparin is to
be seriously considered (2,5,8,9,19). The first step to take in
therapy is the immediate and vigorous treatment of the underlying
disease which provoked DIC, and therefore, a complete understanding
of the mechanisms involved in this underlying condition contributes
to a rational and logical treatment of the patient. In addition
to the treatment of the causative disease, the treatment of DIC
is based on the use of anticoagulants, generally heparin, in order
to block the phenomenon of intravascular coagulation, and the use

	Remission			Failure		
AT III Level	No. of Patients	A	B	No. of Patients	A	B
<50	5	5	0	13	9	4
>50	9	8	1	1	0	1

A) low dose heparin
B) without heparin

Fig. 3. Correlation with AT III and treatment response
         in APL made in our studies.

of transfusions with hemoderivates and platelet concentrates in
order to correct the defect in hemostasis, which is secondary to
the massive consumption (1,3-5,10,19).

It is clear that before proceeding with specific therapy
several factors must be evaluated: the level of fibrinogen,
platelets, and coagulation factors, and whether or not the patient
is hemorrhaging. In case of a self-limiting DIC, one can avoid
treatment if the patient is not hemorrhaging, and confine therapy
to "wait and see," while continuously monitoring hemostastic equilibrium. If the patient is hemorrhaging and has an active DIC,
it is necessary to treat with replacement therapy with heparin
"coverage." Heparin therapy is particularly useful in the initial
phase of those forms of DIC where fibrin deposition causes a
multiregional ischemo-necrotic phenomenon with venous thromboembolism.

The mechanism of the heparin-catalytic effect towards the
AT III-thrombin reaction proceeds via heparin-thrombin complex.
AT III depletion during heparin therapy could be avoided by using
very low heparin levels which could be sufficient to catalyze
thrombin inactivation, but which would not allow AT III-heparin
complexes to form with undesired enzyme inhibition (e.g., plasmin)
(4,13,15,19). Diminution of AT III to less than 70% of normal
level causes an increased thrombotic tendency and a considerable

decrease of heparin effect on coagulation. Hence, the efficacy of
heparin therapy is warranted only by sufficient reserves and
continuous biosynthesis of AT III (Fig. 3). Low AT III levels in DIC
reflect augmented consumption and insufficient protein synthesis.
Substitution of ATIII is possible with the use of fresh plasma and
a commercially available ATIII concentrate (1,3,5).

In a previous paper, we reported our experience in acute
promyelocytic leukemia (APL), a distinct entity in the group of
acute non-lymphoid leukemia because of the frequent occurrence of
DIC (5). Our 28 patients with APL complicated by DIC received an
induction treatment with adriamycin at a dose of 30 mg/sqm daily
for four days and the course was repeated a maximum of three times.
A first group of 22 patients received heparin at a dose of 200 U/Kg
daily, while the other patients were only treated with supportive
therapy (platelet concentrates, fibrinogen, fresh plasma). Only
one patient survived among the group not treated with heparin,
while 13 out of 22 patients of the other group had complete remis-
sion. This suggests that heparin is helpful during remission-
induction in acute promyelocytic leukemia (Fig. 4).

Fundamental therapeutic measures are those directed towards
controlling shock, cardiorespiratory involvement, and acute renal
insufficiency which often complicate the acute defibrinization
syndrome.

In the chronic and subacute forms, therapy varies according
to the underlying condition, the presence or absence of hemorrhage,
and/or signs of thrombosis (6). Even in these forms, treatment of
the underlying illness rates first with respect to therapy, and
often leads to resolution of the syndrome. A tendency towards clot
formation justifies the initiation of heparin therapy, but replacement should be avoided because it may lead to the worsening of the
clinical picture and evolution of the DIC syndrome. In these cases,
the use of fibrinolytic inhibitors is contraindicated because of
disastrous consequences which they may provoke. On the other hand,
the use of urokinase and streptokinase has been suggested in case
of renal involvement in hemolytic-uremic syndrome. The rapid
initiation together with the personalization of therapy represent
the fundamental concepts for a successful outcome in the face of
this syndrome sometimes fulminant, and often uneffected by delayed
treatment.

	No. of Patients	Sex M	Sex F	Age (years)	Remission	Failure
Group A Heparin 200 U/Kg/daily	22	13	9	18-63	13 (59.1%)	7 9 (40.9%)
Group B Without Heparin	6	3	3	32-67	1 (16.6%)	5 (83.4%)

Fig. 4. Results of heparin treatment in patients with APL.

REFERENCES

1. E. Anker, U. Abilgaard, R. Andersen, M. Fagerol, and G. Bjune,
    Low antithrombin in severe disease: consumption or decreased
    synthesis? Scand. J. Haemtol., 30(Suppl. 39):59-63 (1983).
2. J. Bernard, Acute promyelocytic leukemia, Proc. 3rd Interna-
    tional Symposium on Therapy of Acute Leukemias, Rome,
    December 11-14 (1982).
3. B. Blauhut, S. Necek, H. Vinazzer, and H. Bergmann, Substi-
    stitution therapy with ATIII concentrate in shock and DIC,
    Thromb. Res., 27:271 (1982).
4. B. Bonen, P. Sie, C. Caranobe, and J. Guittard, Plasma anti-
    thrombins during heparin therapy, Thromb. Res., 27:123
    (1982).
5. P. Citarrella, R. Perricone, V. Abbadessa, I. Carreca, M.
    Pecorella, P. Di Marco, and A. Cajozzo, Initial thera-
    peutical management of acute promyelocytic leukaemia (APL):
    results of 12 cases, Haematologica, 64:339 (1979).
6. C.J. Diskin and A.B. Weitberg, Treatment of chronic intra-
    vascular coagulation syndrome, Arch. Intern. Med., 140:
    263 (1980).
7. M.B. Donati and E. Vincenzi, Tumori ed emostasi, Atti VIII
    Corso Nazionale di Aggiornamento in Ematologia Clinica,
    Brisighella, April 22-25, p. 511 (1982).
8. D.I. Feinstein, Diagnosis and management of disseminated
    intravascular coagulation. The role of heparin therapy,
    Blood, 60:284 (1981).
9. G. Gamba, G. Grignani, M. Biancardi, D. Geraldi, and E. Ascari,
    Acute leukemias and DIC: effectiveness of treatment with
    low doses of heparin, Haematologica, 64:322 (1979).
10. G.M. Gandolfo and M.V. Torresi, "Antitrombina III," Piccin,
    Padova (1981).
11. L. Gugliotti, S. Viganò, A. Guarini, B. Timoncini, P.M.
    Mannucci, and S. Tura, Disseminated intravascular coagu-
    lation and fibrinopeptide A in adult acute myeloid leukemia,
    Proc. 3rd International Symposium on Therapy of Acute
    Leukemias, Rome, December 11-14 (1982).
12. H.S. Jacobs, P.R. Craddock, D.E. Hammerschmidt, and C.F.
    Maldow, Complement-induced granulocyte aggregation,
    N. Engl. J. Med., 302:789 (1980).
13. I. Majolino, G. Nalli, P. Panizza, L. Cervini, and E. Ascari,
    Antitrombina III: fisiopatologia e clinica, Haematologica,
    64:322 (1979).

14. P.M. Mannucci, Effect on blood coagulation of massive intra-
    vascular haemolysis, Blood, 33:207 (1969).
15. E. Marciniak and J.P. Gockerman, Heparin-induced decrease in
    circulating antithrombin III, Lancet, ii:581 (1977).
16. G.G. Neri Serneri, G.F. Gensini, and R. Abbate, High molecular
    weight fibrinogen complexes in the assessment of hyper-
    coagulability, in: "Haemostasis and Thrombosis," G.G.
    Neri Serneri and C.R.M. Prentice, eds., Academic Press,
    Mondon, p. 123 (1979).
17. M.C. Petti, S. Amadori, G. Avvisati, M. Baccarani, A. Cafolla,
    D. Ruggero, S. Tura, and F. Mandelli, Acute promyelocytic
    leukemia. A collaborative study of 62 patients, Proc.
    3rd International Symposium on Therapy of Acute Leukemias,
    Rome, December 11-14 (1982).
18. T. Siegal, U. Seligsohn, E. Aghai, and M. Modan, Clinical and
    laboratory aspects of disseminated intravascular coagulation
    (DIC), Thromb. Haemost., 39:122 (1978).
19. G.F. Smith and J.L. Sundboom, Heparin and protease inhibition
    II. The role of heparin in the ATIII inactivation of
    thrombin, plasmin, and trypsin, Thromb. Res., 22:115-133
    (1981).
20. J.A. Spero, J.H. Lewis, and U. Hasiba, Disseminated intra-
    vascular coagulation, Thromb. Haemost., 38:33 (1980).
21. J.A. Van Mourik, The significance of fibrinopeptide A (FPA) in
    the diagnosis of low grade intravascular coagulation and
    venous thromboembolism, Haematologica, 66:259 (1981).
22. M.M. Wintrobe, G.R. Lee, D.R. Boggs, T.C. Bithell, J. Foerster,
    J.W. Athens, and J.N. Lukens, Acquired coagulation dis-
    orders, in: "Clinical Haematology," Lea and Febiger,
    Philadelphia, p. 1213 (1981).

β-TG AND PF$_4$ IN THE DIAGNOSIS OF THROMBOSIS

V. Abbadessa, P. Citarrella,
R. Perricone, and P. Di Marco

Patologia Medica I
Piazza delle Cliniche 2
Palermo, Italy

In vitro studies of platelet aggregability have been useful
in the evaluation of platelet function disorders and in monitoring
pharmacological agents. However, these studies are not valid if
applied to thromboembolic disorders. Since the application of these
methodologies is limited, the dosage of platelet release material
has been suggested because they can better demonstrate platelet
activation in vivo.

Attention has been focused on platelet proteins released by
α-granules, in particular β-TG and PF$_4$ that are presently measured
using radioimmunologic methods.

β-TG is an abudant platelet protein that makes us about 2.5%
of the dry weight of platelets. 70% of β-TG is released during
the release reaction. It is secreted as a tetramere of four
identical subunits with a molecular weight of 8.850 (Begg et al.,
1978) so that its entire molecular weight is 35.400. β-TG seems
to be the product of proteolytic breakdown of low-affinity PF$_4$
(LA-PF$_4$), which results in the tetrapeptide cleavage from the N-
terminal of each low-affinity PF$_4$ subunit. Niewiarowski et al.
(1980) have shown that the substances secreted by platelet initially contain 80% of LA-PF$_4$ that is subsequently hydrolyzed by a
protease of platelet origin functioning at 37°C. Plasmin and
trypsin are also able to convert LA-PF$_4$ into β-TG.

PF_4 is also a tetramere. It is secreted as a complex composed of two tetrameres and two proteoglycan molecules with carrier function with an entire molecular weight of about 350,000 (Kaser-Glanzmann et al., 1972; Moore et al., 1975). The complex secreted by platelets seems to disassociate in plasma. The molecular weight of the PF_4 tetramere is 30,800. Like β-TG, it is composed of four identical subunits with a molecular weight of 7,700.

Almost the same content of both proteins is found in platelets (PF_4: 18 μmg/10^9 platelets; β-TG: 17.7 μmg/10^9 platelets) while only trace amounts are found in other tissues (brain,liver) (Ludlam and Anderton, 1977). It is interesting to note that the highest concentration in the tissues studies is found in the spleen which is a reflection of the catheteric function performed by this organ. Dawes et al. (1978) studied β-TG and PF_4 clearance when 200 ml of serum containing both proteins were infused into healthy volunteers. In this experimental study, PF_4 clearance was so rapid that it was impossible to determine $T\frac{1}{2}$, while β-TG clearance showed a $T\frac{1}{2}$ of about 100 minutes.

In Rhesus monkeys Musial et al. (1980) showed that the clearance of circulating PF_4 had an exponential curve with a fast and slow PF_4 clearance rate of 2.1 and 7 minutes respectively. The initial rapid PF_4 clearance is mostly due to the intense endothelial cell captation. This was shown in vitro using human endothelial cell cultures and human PF_4 (Busch et al., 1980), and in vivo by the observation that heparin administration in healthy subjects drammatically and rapidly increases PF_4 plasma rate.

Guzzo et al. (1980) showed that β-TG is metabolized by the kidney and that in patients with renal failure plasma levels of β-TG increase in direct relation to serum creatinine (Ludlam and Anderton, 1977; Depperman et al., 1980) while PF_4 is not affected by the deficiency of renal function.

Briefly, PF_4 and β-TG are found in comparable quantities in platelets and are released in the same quantity in vitro (Kaplan, 1978,1979). However, they have a different plasma half-life, since they are removed from the circulation in different ways and with different characteristics. Evaluation of β-TG and PF_4 plasma levels requires careful procedures so as to avoid errors and artifacts during the withdrawal phase as well as in the dosage samples.

The samples are drawn with a pre-cooled syringe at 4°C and are immediately added to an anticoagulant mixture (Edimburgh mixture) composed of EDTA, theophylline, and PGE_1 at 0°C. Most of the kits on the market do not contain PGE_1, even though its use is recommended in order to avoid the release of β-TG and PF_4 in vitro during preparation of platelet-poor plasma (PPP) (Ludlam and Anderton, 1977; Franchi et al., 1980; Randi et al., 1981). β-TG plasma concentrations are made of the sum of two components: a) real β-TG concentration in vivo; b) β-TG released during the withdrawal phase and preparation of PPP.

However, in a study which included diabetics and normal subjects, the Edimburgh group (Borsey et al., 1980), using two diverse methods of PPP preparation and two different anticoagulants (with and without PGE_1), showed that these variants have different independent effects on the measuring values, and that these effects are moreover independent of the type of patient involved. In other words, the use of either anticoagulant does not modify the results. The variables act in an equal manner in patients as well as in controls.

Several authors have suggested (Nossel et al., 1979; Musial et al., 1980; Kaplan and Owen, 1981) the simultaneous dosage of the two proteins for the study of platelet activation in vivo. The normal relation between β-TG and PF_4, suggested at about three, would increase if the increased β-TG values correspond to a real increase of the release in vivo, while an opposite tendency may be seen in cases where the release of the two proteins occurs in vitro. Finally, it must be noted that a correlation between β-TG and the number of platelets was not found when these are between 100,000 and 300,000/mmc (Ludlam and Anderton, 1977). On the other hand, a significant increase was observed with an age increase (after the 4th decade of life), and lower values were found during the fertile age with respect to the corresponding age in men.

Numerous clinical studies have shown that many disorders such as thromboembolic disorders are associated with platelet activation in vivo. Numerous observations indicate an increase in β-TG concentrations in diabetes mellitus. This increase seems to be more marked in patients with complications such as retinopathy, neuropathy, and vasculopathy which are independent of the type of diabetes, of the disease duration, and of treatment (insulin

Table 1. β–TG and PF$_4$ Behavior in 26 Patients with
          Diabetic Retinopathy

	Controls	Patients	Significance
β–TG	29.6 + 19.4	90.3 + 54.2	p<0.01
PF$_4$	11.5 + 7.2	39.8 + 29.3	p<0.01

and oral antidiabetics). Conversely, it seems to be related to
the cholesterol and triglyceride serum levels. Our experience
with a group of diabetic retinopathics confirm the above–mentioned
results (Table 1).

     In cerebral vasculopathies, PF$_4$ concentration is in relation
to the severity and the extent of cerebral damage. In fact,
increased values are found in ictus (Levine et al., 1981). In
unstable angina pectoris it has been shown that the plasma levels
of the two proteins are correlated with episodes of angina, since
increased values are found only within four hours after the episode.
During the intervals, the levels are within normal limits (Kaplan,
1978). In patients with heart valves the increased amounts of
β–TG seem to be in relation to the type and number of artificial
valves (Pumphrey and Dawes, 1981). Increased levels of β–TG and
PF$_4$ have also been found in acute myocardial infarction, and
moreover Handin et al. (1978) stated that we cannot diagnose
infarction when low PF$_4$ plasma concentrations are found in the
beginning of a clinical picture characterized by precordial pain.

     Can these tests distinguish patients or groups of patients that
are likely candidates for thrombosis? Is there any evidence
regarding their predictive value for thrombosis?

     It is extremely difficult to answer these questions, but a
partial response derives from a clinical trial using drugs which
inhibit platelet function. ARIS (Antural Reinfarction Italian
Study) also tried to perform a prospective study in order to
evaluate if several hemostatic parameters may have a predictive
value and give valid information regarding the selection of patients
at coronary thrombotic risk. In this study, significant increases

of β-TG and PF$_4$ levels were found, but similarly to other hemostatic parameters, no significant correlations were found with the clinical endpoints, so that it was impossible to give them a definite prognostic value. Conversely, the patients with clinical endpoints were not characterized by an incidence of hemostatic anomalies of greater severity with respect to the patients that did not present an endpoint regarding the parameters alone as well as the classical risk factors (Cortellaro et al., 1981).

We can presently state that β-TG and PF$_4$ concentration determination, if accurately performed, may be an index of the rate of platelet activation in vivo which confirms that thrombosis is a multifactorial phenomenon. Therefore, it is still impossible to diagnose from a single test regardless of its accuracy.

REFERENCES

Begg, G.S., Pepper, D.S., Chesterman, C.N., and Morgan, F.J., 1978, Complete covalent structure of human β-thromboglobulin, Biochemistry, 17:1739.

Borsey, D.Q., Dawes, J., Frasher, D.M., Prowse, C.V., Elton, R.A., and Clarcke, B.F., 1980, Plasma β-thromboglobulin in diabetes mellitus, Diabetologia, 18:353.

Busch, C., Dawes, J., Pepper, D.S., and Wasteson, A., 1980, Binding of platelet factor 4 to cultured human endothelial cells, Thromb. Res., 19:129.

Cortellaro, M., Boschetti, C., Beggi, P., and Polli, E.E., 1981, Piastrine e cardiopatia ischemica, Haematologica, Atti del 28° Congresso Nazionale della S.I.E., Bari, 28-31 Ottobre, 341.

Dawes, J., Smith, R.C., and Pepper, D.S., The release, distribution, and clearance of human β-thromboglobulin and platelet factor 4, Thromb. Res., 12:851.

Depperman, D., Andrassy, K., Seelig, H., Rizt, E., and Post, D., 1980, β-thromboglobulin is elevated in renal failure without thrombosis, Thromb. Res., 17:63.

Franchi, F., Canciani, M.T., and Mannucci, P.M., 1980, The β-thromboglobulin test, Thromb. Haemostas., 44:107.

Guzzo, J., Niewiarowski, S., Musial, J., Bastl, C., Grossman, R.A., Rao, A.K., Berman, I., and Paul, D., 1980, Secreted platelet proteins with antiheparin and mitogenic activities in chronic renal failure, J. Lab. Clin. Med., 96:102.

Handin, R.I., McDonough, M., and Mesch, M., 1978, Elevation of
        platelet factor 4 in acute myocardial infarction: measurement
        by radioimmunoassay, J. Lab. Clin. Med., 91:340.
Kaplan, K.L., 1978, Proteins secreted by platelets: significance in
        detecting thrombosis, in: "Thrombosis: Animal and Clinical
        Models," H.J. Day, B.A. Molony, and E.E. Nishizawa, eds.,
        Plenum Press, New York.
Kaplan, K.L., 1979, Relationship between platelet release, thrombin
        action, and fibrin formation in vitro, Blood, 54(Suppl. 1):
        284a.
Kaplan, K.L., and J. Owen, 1981, Plasma levels of β-thromboglobulin
        and platelet factor 4 as indeces of platelet activation
        in vivo, Blood, 57:199.
Kaser-Glanzmann, R., Jakabova, M., and Luscher, E.F., 1972,
        Isolation and some properties of the heparin-neutralizing
        factor (PF$_4$) released from human blood platelets, Experentia,
        28:1221.
Levine, P.H., Fisher, M., Fullerton, A.L., Duffy, C.P., and
        Hoogasian, J.J., 1981, Human platelet factor 4: preparation
        from outdated platelet concentrates and application in
        cerebral vascular disease, Am. J. Hematol., 10:375.
Ludlam, C.A., and Anderton, J.L., 1977, Platelet β-thromboglobulin,
        in: "Proceedings of a Conference on Platelet Function
        Testing," H.J. Day, H. Holmsen, and M.B. Zucker, eds.,
        DHEW Publication (NIH), Philadelphia.
Moore, S., Pepper, D.S., and Cash, J.D., 1975, Platelet antiheparin
        activity: the isolation and characterization of platelet
        factor 4 released from thrombin-aggregated washed human
        platelets and its dissociation into subunits and the isolation
        of membrane bound antiheparin activity, Biochem. Biophys.
        Acta, 379:370.
Musial, J., Niewiarowski, S., Edmunds, L.H. Jr., Addonizio, V.P.,
        Nicolau, K.C., and Colman, R.W., 1980, In vivo release and
        turnover of secreted platelet antiheparin proteins in
        Rhesus monkeys (M. Mulatta), Blood, 56:596.
Niewiarowski, S., Walz, D.A., James, P., Rucinsky, B., and Kueppers,
        F., 1980, Identification and separation of secreted platelet
        proteins by isoelectric focusing. Evidence that low
        affinity platelet factor 4 is converted to β-thromboglobulin
        by limited proteolysis, Blood, 55:453.
Nossel, H.L., Wasser, J., Kaplan, K.L., La Gamma, K.S., Yudelman,
        I., and Canfield, R.E., 1979, Sequence of fibrinogen

proteolysis and platelet release after intrauterine infusion of hypertonic saline, J. Clin. Invest., 64:1371.

Pumphrey, C.W., and Dawes, J., 1981, Elevation of plasma β-thromboglobulin in patients with prothesis cardiac valves, Thromb. Res., 22:147.

Randi, M.L., Fabbris, F., Casonato, A., and Girolami, A., 1981, The effect of anticoagulant mixtures on β-TG and PF₄ levels, Thromb. Haemostas., 46:569.

IS PLATELET SURVIVAL A PREDICTIVE TEST

FOR THROMBOSIS?

P. Di Marco, V. Abbadessa,
R. Perricone, and P. Citarrella

Patologia Medica I
Piazza delle Cliniche 2
Palermo, Italy

INTRODUCTION

Interest in the prethrombotic state is mainly based on the expectation that it may be possible to detect persons who will develop a myocardial infarction or a stroke in the near future. However, the numerous definitions of the "prethrombotic state" are often generic. In fact, even though the vascular area involved in the atherosclerotic process is the cause of the interaction of the hematic component with the vascular wall which may be seen in alterations of tests, these modifications are too unclear and aspecific to allow them to have a predictive value.

Thromboembolic disorders are some of the main causes of death in industrialized countries. Therefore, it is obvious that a search for a definition of the prethrombotic state and a methodology able to detect it are also linked to the epidemiological importance of thromboembolic disorders. The prognosis of these disorders may be influenced by an adequate therapy and prophylaxis.

In the 1960's and 1970's many research groups concentrated on the analysis of platelet kinetics using radioisotopic methods (1-4). These studies were based on the hypothesis that the involvement of platelets in thrombotic and thromboembolic processes favors their consumption and results in an increased bone marrow production. However, methods that utilize radioisotopes are very

complex, even though they are very sensitive, and cannot be used
in routine to detect a platelet alteration in vivo.

Before examining the data from the literature regarding the
study of platelet survival, it is useful to ask the following
questions:

- Does a correlation exist between platelet turnover and the
  degree of vascular disease? If so, can it be shown?
- Are the anomalies of platelet kinetics useful as indeces
  for prognosis and therapy?
- Is there a correlation between platelet turnover and other
  functional platelet tests?

DISCUSSION

With the exception of O'Neil and Firkin (5) who found a normal
platelet survival in patients with peripheral sclerotic vasculopathy
studied with Cr51, many other authors found a decreased platelet
survival in atherosclerotic patients: Murphy and Mustard (6) in a
group of 32 atherosclerotic subjects studied with DF32P; Neri
Serneri et al. (7) and Ritchie and Harker (8) in patients with a
history of myocardial infarction using Cr51; Neri Serneri et al.
(9) and Harker and Slichter (10) in patients with chronic cerebrovascular disorder and in subjects with prosthetic heart valves.

Harker and Slichter (11) studied platelet survival to show the
ability of antiaggregating substances (dipyridamole) in interrupting
platelet consumption and arterial thromboembolic events. Steele
et al. (12,13), while studying 68 coronopathic patients, Genton
et al. (14), and our study group (15) all showed a reduced platelet
half-life in about 50% of patients.

More recently Najean et al. (16) studied a group of 73 patients
with varying types of arterial vascular pathology (arteritis,
vasculopathy of the limbs, hemiplegia, myocardial infarction,
patients with heart valves, aorto-femoral bypass). Their ages
ranged from 23 to 83 years. These patients were compared with a
control group with the same age range. Their data do not show an
increase in platelet production. No significant differences were
noted if patients were subdivided according to sex, age, and
severity of the disease. However, a reduced platelet survival

is seen in those with a heart valves and in those who had undergone
bypass surgery in whom the platelet hyperconsumption seemed constant.

CONCLUSION

The data from the literature are often contradictory. However,
a critical review of the cases studies (method, type of patients,
age group, number of patients) shows the existence of a considerable
overlap between patients and controls which does not allow one to
consider platelet half-life a useful test, if evaluated alone.
Moreover, the alteration of platelet survival must be placed in
relation to the type and entity of the lesion. Platelet kinetics
are not able to photograph a dynamic event such as a platelet
consumption linked to spasm, or other causes that behave in a
dynamic manner. Conversely, a continuous stress condition, such
as a heart valve, bypass surgery, or extensive trauma of the
vascular bed, can be documented.

However, alterations of platelet survival are not necessarily
linked to thrombosis. One must also take into consideration that
many stimuli which are not thrombogenic can interact with platelets,
and induce a release reaction as well as a decrease in platelet
survival, so that these alterations are non-specific and not
necessarily linked to thrombosis and therefore cannot be considered
useful from a predictive point of view.

REFERENCES

1.   Y. Najean and N. Ardaillou, The use of 75-Se-Methionine for
     the in vivo study of platelet kinetics, Scand. J. Haematol.,
     6:495 (1969).
2.   L.A. Harker, R. Ross, S.J. Slichter, and G.R. Scott, Homo-
     cystine-induced arteriosclerosis. The role of endothelial
     cell injury and platelet response in its genesis, J. Clin.
     Invest., 58:731 (1976).
3.   J.L. Ritchie and L.A. Harker, Platelet and fibrinogen survival
     in coronary atherosclerosis, Am. J. Cardiol., 39:595 (1977).
4.   A. Cajozzo, P. Di Marco, and G. Licata, Gli isotopi radioattivi
     nella diagnosi della malattia trombotica, Proceedings from
     "Aggregazione Piastrinica in Fisiopatologia e Clinica,"
     Catania, March 11-12 (1977).
5.   B. O'Neil and B. Firkin, Platelet survival studies in
     coagulation disorders, thrombocythemia, and conditions

      associated with atherosclerosis, J. Lab. Clin. Med.,
      64:188 (1964).

6.   E.A. Murphy and J.F. Mustard, Coagulation tests and platelet
      economy in atherosclerotic and control subjects, Circulation,
      25:114 (1962è.

7.   G.G. Neri Serneri, P. Paoletti, E. Silvestrini, G. Gensini,
      G. Masotti, and R. Abbate, La sindrome trombofilica della
      malattia ateromasica: aspetti patogenetici e semeiologici,
      Giorn. Ateroscl., 7:409 (1970).

8.   J.L. Ritchie and L.A. Harker, Platelet, fibrinogen, and
      plasminogen survival in patients with coronary artery
      disease, Circulation, 50(Suppl. III):285 (1974).

9.   G.G. Neri Serneri, E. Silvestrini, G.F. Gensini, and R. Abbate,
      Gensini, Platelet hyperreactivity and decreased platelet
      survival in chronic cerebrovascular patients - chronic
      defibrination syndrome? in: "Platelet Aggregation in
      Pathogenesis of Cerebrovascular Disorders," A. Agnoli
      and C. Fazio, eds., Proc. Round Table Conf., Rome, 1974,
      Springer Verlag, Berlin (1977).

10.   L.A. Harker and S.J. Slichter, Studies of platelet and
      fibrinogen kinetics in patients with prosthetic heart
      valves, N. Engl. J. Med., 283:1302 (1970).

11.   L.A. Harker and S.J. Slichter, Platelet and fibrinogen
      consumption in man, N. Engl. J. Med., 287:999 (1972b).

12.   P. Steele, H.S. Weily, H. Davies, and E. Genton, Platelet
      function studies in coronary artery disease, Circulation,
      48:1194 (1973).

13.   P. Steele, D. Battock, and E. Genton, Effects of clofibrate
      and sulfinpyrazone on platelet survival time in coronary
      artery disease, Circulation, 52:473 (1975).

14.   E. Genton, Cardiac Thromboembolism: evidence for role of
      platelets and value of platelet suppressant therapy,
      Adv. Exp. Med. Biol., 102:227 (1978).

15.   P. Di Marco, R. Perricone, A. Indovina, and A. Cajozzo,
      Sopravvivenza e turnover del fibrinogeno e delle piastrine
      marcate con 75-seleniometionina in vasculopatici sclerotici,
      Proc. Eur. Symp. on Advances in Coagulation, Fibrinolysis,
      Platelet aggregation, and Atherosclerosis, Palermo 6-9
      October (1976).

16.   Y. Najean, E. Dassin, C. Renner, and M. Wacquet, Cinetique
      Plaquettaire au cours des maladies artérielles - artérites,
      prothèses valvulaire et vasculaire: 73 observations, Nouv.
      Press Med., 8:3813 (1979).

CONTRIBUTORS

V. Abbadessa
Patologia Medica I
90100 Palermo Italy

E. Ascari
Clinica Medica I
Pavia, Italy

A. Baserga
Clinica Medica
Ferrara, Italy

T. Barbui
Divisione di Ematologia
Bergamo, Italy

E. Cacciola
Cattedra di Ematologia
Catania, Italy

A. Cajozzo
Patologia Medica I
Palermo, Italy

G. Camussi
Cattedra di Nefrologia
Turin, Italy

N. Ciavarella
Centro Emofilia
Bari, Italy

P. Citarrella
Patologia Medica I
Palermo, Italy

D. Collen
Katholieke Universiteit
Leuven, Belgium

M. Cortellaro
Clinica Medica I
Milan, Italy

P. Di Marco
Patologia Medica I
Palermo, Italy

T. Di Perri
Patologia Medica
Siena Italy

M.B. Donati
Istituto M. Negri
Milan, Italy

J.A. Dormandy
St. James Hospital
London, U.K.

C. Fieschi
Clinica Neurologica
Rome, Italy

F. Gavosto
Patologia Medica I
Turin, Italy

G.F. Gensini
Clinica Medica I
Florence, Italy

L. Gugliotta
Istituto di Ematologia
"Lorenzo e Ariosto Seragnoli"
Bologna, Italy

V.V. Kakkar
King's College Hospital
Medical School
London, U.K.

D.A. Lane
Charing Cross Hospital
London, U.K.

F. Mandelli
Cattedra di Ematologia
Rome, Italy

P.M. Mannucci
Centro per lo studio e la terapia
delle malattie emorragiche e
della trombosi
Milan, Italy

G. Mariani
Cattedra di Ematologia
Rome, Italy

D. Meyer
Institut de Pathologie Cellulaire
Bicêtre, France

J.K. Morris
Queen's Medical Centre
Nottingham, U.K.

D.F. Mosher
University of Wisconsin
Madison, Wisconsin U.S.A.

J.F. Mustard
McMaster University
Hamilton, Canada

G.G. Neri Serneri
Clinica Medica I
Florence, Italy

R. Perricone
Patologia Medica
Palermo, Italy

F. Piovella
Clinica Medica I
Milan, Italy

E. Polli
Clinica Medica I
Milan, Italy

M. Prencipe
Clinica Neurologica
Rome, Italy

R. Ross
University of Washington
Seattle, Washington U.S.A.

N. Semeraro
Istituto di Patologia Generale
Bari, Italy

A. Strano
Clinica Medica I
Palermo, Italy

M. Tesi
Divisione di Angiologia
Arcispedale S.M. Nuova
Florence, Italy

S. Tura
Istituto di Ematologia
"Lorenzo e Ariosto Seragnoli"
Bologna, Italy